Lecture Notes in Morphogenesis

Lecture Notes in Morphogenesis is an interdisciplinary book series which aims to face the questions of emergence, individuation and becoming of forms from several different points of view: those of pure and applied mathematics, of computational algorithms, of biology, of neurophysiology, of cognitive and social structures. The set of questions above concerns all the manifestations of Being, all the manifestations of Life. At the heart of contemporary embryogenesis lies an essential question: How can form emerge from the constant, chaotic flow? How can a sequence of purely informational elements — an a-signifying combination of chemical substances organized in the DNA molecule — evolve into the highly complex and structured forms of the living organism? A similar question can be asked when we deal with the morphogenesis of vision in neural systems and with the creation of evolving synthetic images, since digital technology makes possible the simulation of emergent processes both of living bodies and of visual forms. Finally the very idea that abstract structures of meaning could be captured in terms of morphodynamic evolution opens the door to new models of semiolinguistics, semiotic morphodynamics, and cognitive grammars. An entire heritage of ideas and concepts has to be reconsidered in order to face new and challenging problems: the theoretical framework opened by Goethe with the introduction of the word "Morphogenesis" is developed by D'Arcy Thompson in "On Growth and Form", it is reorganized with new theoretical insights by the classical structuralism of Levi-Strauss and formalized by the dynamical structuralism of René Thom. The introduction of the post-structuralists ideas of individuation (in Gilbert Simondon and Gilles Deleuze) and plasticity of structures builds a bridge to contemporary problems of morphogenesis at a physical, biological, social and transindividual level. The objective of this book series is to provide suitable theoretical and practical tools for describing evolutionary phenomena at the level of Free boundary problems in Mathematics, Embryogenesis, Image Evolution in Visual Perception, Visual Models of Morphogenesis, Neuromathematics, Autonomy and Self-Organization, Morphogenetic Emergence and Individuation, Theoretical Biology, Cognitive Morphodynamics, Cities Evolution, Semiotics, Subjectivation processes, Social movements as well as new frontiers of Aesthetics. To submit a proposal or request further information, please use the PDF Proposal Form or contact directly: *Dr. Thomas Ditzinger (thomas.ditzinger@springer.com)*

Silvia Vizzardelli · Riccardo Panattoni
Editors

Plasticity and Discontinuity

Morphogenesis Beyond Structure

Springer

Editors
Silvia Vizzardelli
Department of Humanities
University of Verona
Verona, Italy

Riccardo Panattoni
Department of Humanities
University of Verona
Verona, Italy

ISSN 2195-1934 ISSN 2195-1942 (electronic)
Lecture Notes in Morphogenesis
ISBN 978-3-032-03118-1 ISBN 978-3-032-03119-8 (eBook)
https://doi.org/10.1007/978-3-032-03119-8

This Springer imprint is published by the registered company Springer Nature Switzerland AG
The registered company address is: Gewerbestrasse 11, 6330 Cham, Switzerland

Preface

A linguistic hesitation, a lexical *impasse*, often signals the imminence of the new. Something presses forward, revealing the inadequacy of the words used up to that moment, because an excess gestures elsewhere. The current debate on morphogenesis and heterogenesis betrays this hesitation, becoming even more urgent. The hesitation arises when we attempt to think together what has traditionally been kept apart: the continuous and the discontinuous, process and event, duration and leap. The difficulty is not merely terminological, but ontological: how to name what happens in the interval, on the threshold, in the fold between a linear conception of becoming and the irruption of discontinuity; between cumulative processes and generative events? It thus becomes clear that the initial lexical impasse is not merely the symptom of a tension awaiting denomination, but the very attempt to hold together—within a single conceptual gesture—what at first appears irreconcilable.

To engage seriously with the question of heterogenesis is to accept the task of becoming *onomaturgists*—name-givers, formers, inventors of concepts. Perhaps feeling unworthy of such an arduous task, we have chosen, in this book, the path of a long backward step into the history of thought, recovering ancient etyma which, if nothing else, have the virtue of disturbing our linguistic habits and, in so doing, of opening the way for novelty. Antiquity as future.

For now, suffice it to say that this return to a classical lexicon serves a function far from secondary: it protects us from reductionism of any kind, for it sustains the tension of ambivalence and is capable of lingering in spaces unmarked by univocal polarizations. If we consider that the most insidious enemy of heterogenesis is precisely reductionism, then perhaps we are on the right path.

The hypothesis we shall attempt to articulate is the following: *in order to grasp something of heterogenesis, we must find an escape route from both the labyrinth of continuity and the labyrinth of discontinuity. That is to say, continuity and discontinuity may, paradoxically, co-implicate each other without collapsing into nonsense.* We might even suggest that the true challenge of contemporary thought lies precisely in the effort to orient itself within this aporetic coexistence.

In an era marked by the proliferation of control apparatuses, knowledge archives, and predictive grammars, thought is called upon to respond not through synthesis or systematization, but by listening to the movements that escape, the folds that elude capture, the tensions that displace themselves. The essays collected here move along this line of inquiry, traversing fields that may appear disparate—biology and psychoanalysis, philosophy of science and aesthetics, linguistics, and ontology—but converge around a central knot: to think the living not as a given entity, but as a transitory and differential figure, the effect of tensions, drifts, fractures, and reversals. The task is not to describe life, but to grasp it in its act of appearance, in the gesture that makes it real while emerging from a groundless space.

There is thus no science of life, unless one considers precisely that which in life eludes all knowledge—that which Leibniz called *petites perceptions*, which for Bergson is duration, intensity, intuition, and which Lacan names jouissance.

From this framework, the volume unfolds along a path that strains classical notions of biology, knowledge, perception, individuality, and subjectivity. It explores human ontogenesis and the infans as a radical threshold of language; the aesthetic foundation of perceptual knowledge in Leibnizian monadology; the emergent singularity of the trait; the *fulguratio* as an unforeseen and generative act; Lacanian biology as a subversion of the Aristotelian paradigm; the intensive calculus and the writing of movement prior to any form; the metastability of the individuation process; and subjectivity as a plastic and discontinuous emergence of meaning rising like an upsurge from the sea—thus urging thought to reformulate its own instruments.

The aim is not to establish a new system, but to displace the position of thought in relation to the living. In this sense, heterogenesis—non-linear, non-causal, non-deductive generation—becomes the cipher uniting theoretical genealogies and experiential practices, even when distant from one another, all engaged in the effort to think that which is born, that which comes into being.

A new biology, a new aesthetics, a new logic—or perhaps a return to those heretical undercurrents always present beneath the surface of Western knowledge—emerge through the essays collected here. On the condition, however, of accepting the risk of discontinuity, rupture—and with them, the very enigma of generation.

In his essay *Ex nihilo*, Federico Leoni—drawing inspiration from Borges, Foucault, and the non-Western logic of Chinese encyclopedias—challenges the model of classification based on fixed traits (the "Porphyrian oak") that underpins Western ontology. He proposes an ichnology: a form of knowledge grounded in the gesture, the trace, the emergent singularity. In this framework, the world is not built by the accumulation of categories but by events that trace fleeting yet decisive differences. Every trait is a threshold, every classification a poetic act, every identity the retroactive effect of an identification that precedes and undoes us.

A similar emphasis on discontinuity is found in Silvia Vizzardelli's essay, where the ancient scholastic concept of *fulguratio* is reactivated as a figure of the creative leap, the moment in which a form appears suddenly as a performative act. Generation, from this perspective, is not the consequence of a process, but the event that grounds the very possibility of a process. Through a trajectory encompassing thinkers such as Leibniz, Konrad Lorenz, Nicolai Hartmann, Georges Bataille, Jacques Derrida, and Jacques Lacan, she outlines a genealogy of discontinuous thought, a current sometimes repressed, subterranean, yet ready to erupt anew as in the case of Leibniz, or to immediately seize the stage and never relinquish it.

In *Fundus Animae*, Domenico Licciardi addresses perception through Leibniz, Herder and Deleuze: at its center lies the function of the *petites perceptions*, pre-cognitive forms that ground sensible experience. *Aisthesis* is thus conceived as a mode of knowledge irreducible to representation—a mathematics of sensibility that operates at the boundary between the continuous and the discrete. Perception, not as data registration, but as active and generative function, capable of constituting worlds from infinitesimal affective variations, appears as a tensional field in which subject and object co-constitute one another.

In *Genus*, Matteo Bonazzi adopts psychoanalysis as a lens through which to refract biology: in particular, Lacan's later *Seminars* become the lever by which to displace the Aristotelian notion of the living. The speaking body, inscribed in language prior to biological individuation, is conceived as the effect of the signifier—a killing of the Thing that paradoxically, becomes the cause of jouissance. Life here is not a substance but a writing; the subject is constituted from a constitutive void, a primordial split; the *point of life* coincides with the enigma of the object, which both generates and disrupts the subject's unity.

Marco Mazzeo's *Infans* dwells on ontogenesis as epistemological scandal: the child is defined by the absence of speech (*infans*), and it is precisely this absence that opens the scene of the human. Childhood is not a transitory phase, but a chronic condition of the species, a threshold revealing the subject's dependence on play, on fiction. From the outset, the human is neither sovereign nor self-possessed: the infantile condition shows that the human is exposure, openness, play, and anticipation. Identity is forged through fictions, language games, and non-intentional acts. The figure of the *infans* also poses a challenge to how we think politics and ethics. Deprived of a voice of its own, the infans interrogates the legitimacy of public speech, the status of representation, and the act of speaking itself.

In *Intensio*, Alessandra Campo develops a profound genealogy of the infinitesimal as a "psychic" category, which renders the interior of movement thinkable. This interior, variously designated as impetus, motion, or momentum, is the cause of the emergence of something—a magnitude, a form, or a being (if infinitesimals and their products—the finite extensities—are ontologized)—from nothing. Traversing Leibniz, Kant, Bergson, and Ruyer, differential calculus is read as the instrument of a new phenomenology: not a description of the world, but an ontology of intensity, in which the sensible datum is produced, not presupposed.

Riccardo Panattoni's *Modus* proposes a modal ontology capable of overcoming the traditional opposition between being and becoming. Through the concepts of field, diagram, pre-individual, and counter-effectuation, inspired by Gilbert Simondon and Gilles Deleuze, the author shows how individuation is never a completed fact, but a continuous, metastable, impersonal process. The Outside—understood as an oceanic field of forces—bends into the Inside of the living, generating momentary singularities, as in breath or the creative act. The diagram, central to this reflection, is not representation but modulation of intensities: a matrix of life, thought, and art. In this perspective, subjectivity emerges as a temporary, fragile, and generative fold within a cosmos that remains radically immanent.

Finally, in *Thalassa*, Elena De Silvestri undertakes a radical gesture: to reread the notion of regression through Ferenczi's lens, detaching it from any evolutionary linearity. Here the unconscious is conceived not as a logical-psychological system, but as a topographical and environmental space. The regression toward the "aquatic life" of Thalassa is not merely nostalgia for the maternal, but a movement toward a non-anthropomorphic animism that dissolves the ego into the rhythms and crises of the inorganic. This "oceanic desire" thus becomes the figure of a regressive plasticity, capable of giving form and meaning where trauma has disrupted sense.

Along these trajectories, knowledge ceases to be a device of mastery and becomes a practice of exposure. Exposure to the formless, the discontinuous, the intensive, the catastrophic. No longer the description of reality, but the experience of a reality that is generated within thought itself—as if knowledge were always an operation on the threshold between the human and the inhuman, between the living and the dead, between the image and the imprint. In this sense, each concept is a trace of the impossible.

Verona, Italy
July 2025

Silvia Vizzardelli
Riccardo Panattoni

Contents

Chapter 1
EX NIHILO

Federico Leoni

Abstract This chapter explores the philosophical and ontological implications behind Jorge Luis Borges' famous "Chinese encyclopaedia," which also inspired Michel Foucault in *The Order of Things.* I propose reading this seemingly absurd classification not as a mere logical paradox, but as a clue to a radically different way of understanding the world—one not grounded in stable and shared orders, but in gestures, traits, and emergent differences that produce fleeting yet meaningful singularities. This "ichnological" vision suggests a universe made not of stable beings, but of traces and gestures, aligning thought with performativity and magic rather than structure and identity. Through references to Borges, Foucault, Descartes, the Bible, and Sino-Japanese linguistics (particularly the role of classifiers), I develop a theory of the trace, in which being is not foundational but produced—an emergence rather than an origin. The essay culminates in a reinterpretation of the Western notion of *creatio ex nihilo*, not as a metaphysical foundation, but as a symptom of the distorted and unresolved experience of pure singularity.

1.1 China is Close

China is not so far, we used to say. And yet China is far away and has remained so, in philosophy. Jorge Luis Borges, in a famous short essay, portrayed a mythical and fable-like China, which is however not as crazy and imaginary than Borges would have us believe, and is necessary now more than ever for a new metaphysics. By focusing a little more closely on some of its more suggestive aspects, we could perhaps learn a few lessons from it.

The Analytical Language of John Wilkins, the title of Borges' short essay, references an obscure German scholar, Franz Kuhn, whose work apparently mentions an even more obscure and, in any case, unspecified Chinese encyclopaedia. This encyclopaedia attempts to classify animals under the following categories: "(a) those that belong to the Emperor, (b) embalmed ones, (c) those that are trained, (d) suckling

F. Leoni (✉)
Università di Verona, Verona, Italy
e-mail: federico.leoni@univr.it

S. Vizzardelli and R. Panattoni (eds.), *Plasticity and Discontinuity*, Lecture Notes in Morphogenesis, https://doi.org/10.1007/978-3-032-03119-8_1

pigs, (e) mermaids, (f) fabulous ones, (g) stray dogs, (h) those that are included in the present classification, (i) those that tremble as if they were mad, (j) innumerable ones; k) those drawn with a very fine camel's-hair brush, (l) etcetera, (m) those that have just broken a flower vase, (n) those that at a distance look like flies" [2, p. 126].

1.2 An Attempt to Neutralise the Chinese Encyclopaedia

Over the twentieth century, Borges' essay enjoyed an illustrious philosophical legacy. As is well-known, Michel Foucault took his cue from it at the beginning of his vast study published as *The Order of Things*. In the introduction to this book, Foucault describes the Chinese encyclopaedia as the "birthplace" of his enquiry. And what he finds, in the "laughter that shakes us when we read it", is something that could equally shake, he says, "our thousand-year practice of the Same and the Other" [7, XV].

He specifies that what that laughter unsettles is the consistency of the space in which those bizarrely classified animals are expected to coexist. The madness or strangeness of the classification does not, in fact, lie in the individual categories. Trained animals exist, as do animals painted with a camelhair brush. Mermaids also exist, at least if we establish that existing includes being present in certain texts, in a certain imagery, and so on. Nor is it true that the madness or strangeness of this classification lies in the paradox of the "others" or "those that are included in the present classification", which Foucault dismisses somewhat quickly as referring to "well-known paradoxes" [7, p. XVII]. It lies, on the contrary, as Foucault repeats in various ways, in the absence of a "common space" in which those beings might live together. In the case of the Chinese encyclopaedia, this space of cohabitation seems to be reduced to the "simple sequence of letters of the alphabet" (Ibid. p. XVI) that indicate the various categories.

This sense of estrangement is Foucault's starting point for his enquiry. It is impossible for us to conceive the universe that the Chinese encyclopaedia displays to us, and thus the use we must make of this impossibility will necessarily be self-critical, autobiographical and auto-genealogical. The Chinese encyclopaedia appears to make us face a "limit which is our own" (Ibid. p. XV). Might not we, who find our own way of classifying beings so familiar, be in turn be strange or exotic, prisoners of the false evidence provided by our own equally fanciful criteria, blinded by an obviousness that we should instead put into question? What Foucault prefigures is, put briefly, an entirely negative programme. It seems to be a matter of using parentheses, and suspending the course of our habitual mode of constituting the world. There are echoes of old Husserl in Foucault's gesture. Even more so, echoes of old Kant. What is the a priori of this judgement of ours on the world, of this habit we have of subdividing the universe according to one system of identities and differences, rather than using other systems of identities and differences?

1.3 A Note on Sino-Japanese Classifiers

Could we, instead, make an affirmative use of the Chinese encyclopaedia? Could we adopt it, searching for its spirit if not its letter, and learn something from its madness? Could we go so far as to imagine that what seems like madness to our reason is simply a different reason, and to which our reason could easily appear to be a particular kind of madness?

Scholars have pored over the few traces Borges offers of his encyclopaedia. Borges says he drew on the writings of a certain Franz Kuhn, as mentioned above, an English translator of Chinese texts who is now virtually unknown. Some have pursued this line of research [4, pp. 103–115], arriving at Kuhn's writings and the Chinese encyclopaedias he must have consulted. These encyclopaedias actually did exist, and actually were based on criteria that the Western mind truly finds bizarre.

We might also add a broader linguistic consideration thanks to this work that has been done on Borges Chinese sources. The encyclopaedic criteria we find so bizarre are rooted in a circumstance linked to how the Chinese language functions, as does Japanese; or perhaps to a way of life that is best expressed by how those languages function, and which not by chance surfaces in the encyclopaedias written in those languages.

There are in fact particles in those languages, called classifiers, which are used whenever it is necessary to give a number to a certain object or a certain quantity of objects [10]. They serve to make explicit in advance a kind of broader category, to which those objects we are about to speak belong. Imagine, for example, saying: one umbrella, three guns. Or: one plate, six sheets of paper. We would not use the numeral, "one", "three" or "six" without adding a classifier, that is, a particle that makes it clear that what we are numbering, however, is a long and narrow thing, in the case of the umbrella, or a flat and thin thing, in the case of the plate and the sheet. It is as if we were never dealing with the same "one"; as if it were one thing to say that we are dealing with a single but long and narrow object, and another to say that we are dealing with a single but flat and thin object.

Even more peculiar is the list of things that share the same classifier. One of them, used for flat and thin things, applies to plates, sheets of paper, photos, rafts and garments. Another, used for long and narrow things, applies to rivers, roads, ties, bottles, trains, umbrellas and films. There are many others, dozens of them, and they all function in the same way, all referring to sets of objects that Europeans would never put together. Doesn't the world that opens up before us seem to come straight out of Borges? Wouldn't we tend to put ties and garments together, and in another sphere bring together means of transport, for example rafts and trains? And wouldn't we tend to avoid putting together films and ties, or plates and photos?

1.4 The Magical World

A path opens up to us. We immediately understand that it all depends on the trait we want to highlight from time to time, on which trait concerns us most at a given moment or in a certain context. The decisive thing, in short, is the magic moment when a trait is identified. Then the whole classification follows.

The word magic is far from being merely suggestive. It denotes an act or event that does not abide by the laws of the ordinary world, and yet leads to effects on the ordinary world. The top hat was empty, and a rabbit sprang out of it. Now that the rabbit has leapt out, that rabbit is part of the world, and the world has become a world in which rabbits can pop out of a hat, at least if the hat in question is a top hat. The old world did not contain the possibility of that given event, but that given event, once it has happened, becomes part of a new world. And in the new world this event is part of its ordinary functioning, i.e. its possibilities and the chance those possibilities have of realisation.

Let us return, then, to the question of the trait, and the magical moment in which a certain trait comes to light, with its set of possibilities and realisations of those possibilities, both of which, as we will gradually understand, are made possible by the emergence of the trait. It's as if the trait made a certain possibility possible, or introduced a certain power into the world, and also the eventual actualisation of that possibility, the eventual realisation of that power. Flat and thin things, says the Chinese encyclopaedia. Or long and narrow things. Or things painted with very fine brushes, says the Chinese encyclopaedia. Or things that look like flies from a distance. Who knows, maybe it would be interesting, for some purpose we cannot imagine, to wander through the vast world gathering all the things that look like flies from a distance, and resolutely distinguishing them from all those that do not.

After all, how on earth can we be so sure that the decisive trait is the one whereby certain animals lay eggs while others give birth to offspring that bypass the egg, unless we consider the placenta to be a completely internal egg, soft rather than hard, portable rather than layable? Is it not based on having chosen some other trait, that we then chose viviparity as another relevant trait, making viviparity/not viviparity an indispensable distinction? This infinite regress allows anyone to say that there are good reasons for saying what he is saying, but those good reasons that precede our current way of saying actually come after our current way of saying, and our current way of saying brings them to light and arranges them upstream, much more than they could give a foundation to our current way of saying by arranging it in downstream.

1.5 Holding the Trait

In this sense, the decisive moment is the one, and only the one in which a certain trait takes on prominence, and as a consequence of it taking on prominence, a more general thing also takes on prominence, a category corresponding to that trait; for example, all things that from a distance look like flies. Once again as a consequence of this, a certain particular thing or a certain series of particular things are drawn into the orbit of that general thing, becoming its representatives. The trait, thus, is neither a particular nor a universal, but a power that draws a line through a certain field by relating the two emerging regions, that draws a line through the world by subdividing it into certain universals and the particulars of those certain universals, which precisely for this common origin will remain related.

If every journey begins by leaving home, we can now understand quite clearly what we must take leave of, if we are to learn to think in accordance with the indications of Borges' encyclopaedia. Under (a) we find animals that belong to the emperor. Under (b), therefore, we should find animals that do not belong to the emperor, but to someone else. At that point, letter (c) should indicate a division within the previous ones, for instance the empress, or the court dignitaries. What happens, in short, in a Western encyclopaedia, is that from time to time we choose a trait that for some reason interests us, for example the animals belonging to the emperor, then we elevate that trait to our point of view on the set of all animals, and begin to form an initial set of animals that possess this trait, only to discover that a second set has been produced, including the animals that do not possess this trait.

A Western encyclopaedia holds the trait steady, long enough for this series of reasonings and implications to unfold. A Chinese encyclopaedia, it seems, functions in a very different way. It evokes a kind of magical world, which continually invents new categories and thus newly categorised individuals, if we were to express it in cognitive terms. A world that continually invents new possibilities and therefore new actualisations of those possibilities, if we were to express the same thing in so-called ontological terms. But we shall see that referring to ontology does not do justice to the Chinese encyclopaedia and the world it announces, and that the more ontology becomes thin, the less being there is, and the thinner the difference becomes between being and thinking, between ontology and gnoseology, between things and what we say about them. Are we really facing the animals of the encyclopaedia frontally, like in a zoo? Or are we part of the zoo, and therefore not in front of the encyclopaedia, but inside it?

1.6 A Digression on the Negative

We might actually note that within our occidental encyclopaedia, having isolated a given trait, the animals that possess it are in fact first and foremost imagined as animals that are deprived of it, and then receive it somehow positively, but also contingently.

The price to be paid for this positivity, deep down, is to deal with a certain fundamental negativity. The trait in question may belong to them or not; after all, this is the sense in which we speak of possession or property. I do not own my hand, to speak properly, while I own the umbrella I hold with my hand. That is to say, I do not possess something I am one with, I only possess something that I could lose, and which precisely for this reason I can, under certain conditions, possess. In reality, the set of animals lacking that trait is the set that guides the classification, even if this set seems to manifest itself only secondarily, while the set of animals possessing the trait, even though it seems to manifest itself firstly, depends on the other. What is produced, in other words, as a result of this desire to hold the trait steady, is an aprioristic privilege assigned to the negation of the trait, to the deprivation of the trait, which becomes the paradoxical condition of possibility of the affirmation of the trait or the possession of the trait.

At any rate, at this point we repeat the same operation on the second set, the one with animals lacking the trait of belonging to the emperor. If those animals do not belong to the emperor, might they perhaps belong to the empress? In short, as we were saying earlier, we Westerners, we Porphyrians hold that trait steady. We think that the trait is objective, and that the trait is stable. We think that the trait is independent of our whims, that it belongs to being, and that any trait, if we hold it steady and follow it from one level to the next, from one branch to the next, will take us as far as the safe haven that is being, as far as that genus of all genera, that genus so general that it is not even a genus. We think that it is possible to exhaust the universe, precisely because we hold that trait steady; we think that by proceeding from one trait to the next we can arrive at something that doesn't depend on any trait, something that is completely devoid of traits. This is what ensues from our desire or our need to hold the trait steady, and this is the true root of the Western encyclopaedia. The root of the Western encyclopaedia, in other words, is not being, but the will for there to be being. And as we have already seen, the even truer root of this true root is nothingness, the negation of being; it is the vision of being as something that arises in a scenario in which it is imagined to be absent.

1.7 The Porphyrian Tree

This result is what is known in the West as the Porphyrian tree, since it was Porphyry who first theorised it in his *Isagoge*.

"Porphyry," writes Umberto Eco, "conceived a single tree of substances, whereas one can assume that Aristotle would have been more flexible in imagining many more trees, perhaps complementary to each other, and depending on the type of problem involving definitions he had to solve" [6, p. 342].

Be this as it may, when commenting on Aristotle's *Categories*, and only indirectly on the *Prior Analytics*, Porphyry drew up a single tree of substances, and we are the grandchildren of this decision. As soon as we glance at an encyclopaedia that speaks of animals, we immediately think that not only the set of animals that belong to the emperor is a subset of animals, but that animals are also a subset of something larger, for example that of living beings, which also includes for example plants. Then we think that living beings are also a subset of something even broader, for instance the set of beings that simply have the property of being, or rather being in general, that simply is in general.

Umberto Eco observed with his usual irony that Porphyry "apparently intended, we do not know how sincerely, to leave aside the question of whether genera and species exist in themselves or are mere conceptions of the mind" (Ibid., p. 341). Porphyry, put briefly, doesn't seem to be particularly interested in whether beings structure themselves according to the dictates of the tree or whether our categories proceed by negations and bifurcations. The fact remains that in an initial direction, the tree will lead us to postulate, concerning both things and concepts, the existence of gradually more sizeable branches, up to the enormous trunk representing the being that unites all beings in their pure and simple being of beings. But the tree also moves in the opposite direction, and in the opposite direction will therefore lead us to postulate that the branches perpetually becoming thinner, and ultimately fray into a vast array of twigs consisting of a single, increasingly specific attribute. Animals that belong to the emperor, but only if that emperor who is dressed in dark green and is considered on a Monday morning. Beings, individual substances, things in their singularity, will consist, in general, in a kind of geometric place, which can be circumscribed by following a certain number of divisions of the trunk, along a line that moves between the extremely general being and the extremely particular attribute, and stopping somewhere along the tree, at a certain intermediate node.

But this geometric place is clearly unstable to the utmost degree. To remain there is ultimately impossible. This is the price to be paid for wanting to hold the trait steady. The tree will trace out a twofold movement, more than a twofold direction. This twofold movement will shoot simultaneously towards the summit and towards the root, towards the absolute universality of being and towards the absolute particularity of beings. Every singularity will be divided and sent back with a sort of hasty ferocity towards those opposite extremes. They alone will seem endowed with some truth or consistency. Singularity will somehow be liquidated, nothing more than the intersection between a being that is always a little wider than singularity, and a specification that is always a little narrower. And even of that broader being and that narrower specification, we will have to say that they are in fact nothing more than an intersection between an even broader region, tending towards totality, and an even narrower specification, tending towards the imperceptible. Ultimately, there will be nothing left of singularity. The geometric place that should belong to it will always be irreparably empty.

1.8 On Momentary Substances

The Chinese encyclopaedia proceeds differently. After identifying the animals belonging to the emperor, the encyclopaedist immediately loses all interest in examining other animals from the point of view of belonging to anyone whosoever. There is a magnificent discontinuity in his outlook. He jumps to an entirely different category, for example embalmed animals. And then he jumps again. Animals that are trained, suckling pigs, and so on.

The encyclopaedist, in other words, changes the trait at each division. Indeed, he seems to shun the very principle of division, of a division between having that trait and not having that trait, between belonging to the emperor and not belonging to the emperor. The encyclopaedist enunciates a set of animals having that trait, and because he changes trait immediately afterwards, immediately moving on to something else, he never lets us find, alongside the animals that have that trait, those that do not. We find other animals having a different trait, if anything, a completely different trait that was not mentioned previously. The result is that the Chinese encyclopaedia only ever puts positive sets in front of us, never negative sets. It speaks of a universe that is full, oblivious of lack, because it is through and through unstable and fluctuating; it is so lacking that it completely disregards the possibility of solidifying any possible lack within appropriate boxes. Lack is everywhere, so lack is lacking everywhere. What is there is there every time, everything that is there, everything that cannot not be there.

Let us read Borges' encyclopaedia in more detail, however. Animals that belong to the emperor, animals that look like flies from a distance. It is not that there are only differences, with no genre. It is not that there are only species, lacking some broader category, situated behind them or above them. It is not that there is no subdivision, simply because there is no lack, because there is no negation. We find animals that belong to the emperor, or animals that look like flies from a distance. It is as though, each time, some division was operating, but that division was momentary. We might try to re-read in this way the categories we have already mentioned so many times: animals, but which belong to the emperor. Or else: animals, but which look like flies from a distance. True, we immediately move on to something else, meaning that the branch on which we are now situated does not, so to speak, have time to generate a negative ramification: animals, but which do not belong to the emperor; animals, but which from a distance do not look like flies. And yet, it is equally true that a kind of internal ramification is produced, something like a doubling instead of a division, an incipient bifurcation that does not become a splitting but remains on the verge of being produced. This is the meaning of the "but" that we have tried to introduce to make this aspect explicit.

Put differently, it is as if a particularity had produced its own generality, but that generality lasted only the time of its particularity. As if it was not at all a more general past, to the point of presenting itself as absolute. As if that briefly surfacing generality did not manage to offer a solid ground within which to carve out different species, further species, and negative species, in particular. Rather, we have an instable area that simultaneously generates its substance and its accident, its matter and its form, and then erases itself. And the game begins again, elsewhere. Are we dealing with a universe with no common space, to use Foucault's expression? Yes and no. This is a universe in which there are common spaces, but only momentary ones. It is a universe in which a certain trait emerges, traces a brief common space, traces a space of its own, but a space that is all that there is, all that it is possible to trace. It is a universe of events that simultaneously delineate the being of that entity, and the entity of that being, and immediately disappear. Immediately, or better said, almost immediately. Not without having traced, as we were saying, the substance of that accident and the accident of that substance, the matter of that form and the form of that matter. Not without that event being distributed on the twofold and momentary plane of a quasi-accident and a quasi-substance. This gives rise, so to speak, to momentary substances and momentary accidents, or to accident-substances that seem to emerge from nothingness only to plummet back into nothingness.

1.9 The Phantasm of *Ex Nihilo*

Perhaps the strange idea of the creation *ex nihilo* of things and the universe, which periodically emerges in Western culture, obsessed as it is with the root of the Porfirian tree and its promise of consistency, has to do precisely with the need to think about the singularity of these accident-substances that are so radically devoid of history and projection, so able to elude any form of memory and desire.

We might recall, for example, the creation that the biblical god enacted against the background of a chaos close to non-being: "And the earth was without form, and void; and darkness was upon the face of the deep" [11, 1]. (Or, we might think of the re-creation of the universe and every substance contained within it, which the Cartesian god enacts "almost at every instant" to produce what we perceive as the continuity of the universe, along with the persistence of every single substance, including our own [5, pp.48–49], [8].

What does this phantasm of creation *ex nihilo* speak of? Why does it return so constantly in Western culture? Perhaps the phantasm stems from the need to express, within the language and sensibility of Greek-Jewish-Christian culture, that scandalous vision or experience that is the "Chinese" vision or experience of singularity. And yet, that vision or experience of singularity comes to us in a version, that of the creation *ex nihilo*, which is both deformed and traumatic. Deformed, because it is filtered through the wrong words, words that could not have been abandoned, probably. Traumatic, because only a kind of excess seems to allow this unlocalizable, almost informulable vision to shake the Porfirian tree out of its tragic immobility.

The fact remains that this vision and this experience reach us. They reach us precisely in the deformed form of the *ex nihilo*, I would almost say in the symptomatic form of the *ex nihilo*. And to those who know how to listen to it, it speaks solely of discontinuity. In its own way, it sweeps away the illusion of the matter of a universe believed to be given since the dawn of time, of the uninterrupted continuity of a causality believed to run along a one single line of time always homogeneous to itself. It sweeps away the illusion of a common space. Yet it does not speak of the simple absence of common space. It speaks, if anything, of a momentariness of the common, an occasionality of time, an eventuality of space. Here we must proceed quickly, and roughly. The West imagines a creation *ex nihilo* of things and objects as we know them, of certain combinations of matters and forms that are actually well known to us. It does not imagine that, underneath this initial semantic layer, the image of *ex nihilo* hides a second one, far more vertiginous. This second layer speaks of the creation of both matter and form, of their simultaneous creation by a third element, which as far as he is concerned is neither matter nor form, a third element that we might call gesture.

It is clear, furthermore, that creation *ex nihilo*, both in its Hebrew-Christian version and in its Christian-Cartesian version, is only in partially a creation out of nothing, since it is also a creation enacted in a kind of wasteland, or based on the gesture that extracts matters and forms and things and events from that wasteland. This gesture, in both cases, is attributed to God, as if to signify the foreignness or privilege of the gesture with respect to nothingness, and to discard that nothingness into a subordinate dimension, so that the nothingness will at most be a mirror of the gesture, an inverted figure of it. Thus, if we pay attention to the theology that most radically enquires into the question of nothingness, it is never a question of believing in nothingness, or assuming that nothingness is the ultimate truth of being [9]. Rather, it is always a matter of deciphering nothingness; and upon patient and unprejudiced consideration, nothingness reveals itself to be an anamorphosis of God, a parody of his gesture of creation.

1.10 The Groupist

A short story by Jorge Luis Borges, written with Adolfo Bioy Casares many years after *The Analytical Language of John Wilkins*, is soberly and allusively entitled *The Groupist* [3].

The character who gives the story its title received this nickname because he dedicated his life, we are told, to investigating the banal yet enigmatic problem of how groups, groupings of people or associations of individuals are formed.

Mankind, the Groupist explains, is composed of an infinite number of groups, societies and associations. These are mostly secret societies, he declares. He adds no more, but it is fairly clear that these societies are kept secret not so much from those who are unaware of their existence, but from their own members. What we discover as we keep reading the story confirms this hypothesis. The members do not know they are part of them; they are unaware of the moment at which they are included in them as well as the moment at which are expelled from them. They are unconscious societies, we might say, not secret ones. Rather than being kept secret to the uninitiated, they are as secret to the uninitiated as they are to the initiated.

Borges and Bioy Casares' bizarre hypothesis takes shape along the story. It even becomes clearer when we understand what kind of groups and associations the Groupist is talking about. Some groups are more durable, he says. For example, the one made up of individuals who have a Catalan name, or a name beginning with a certain letter, for example the letter G. Other groups are more transitory, for example the one that brings together all those who at this moment, in Brazil or Africa, are smelling the flower of a jasmine. We might add, even if Borges and Bioy do not dwell on this possibility, which is in any case entirely part of their discourse, that the groups we are talking about partially overlap with other groups, or branch off, subdividing into other groups, and then merging, then disjoining again but along other lines of differentiation. For example, among those currently suffering a coughing attack, some are also wearing slippers. The two groups converge, at least briefly, for as long as the cough lasts and until the slippers give way to shoes, or the idea of walking barefoot on the front lawn. The smallest gesture, the Groupist concludes, such as lighting or extinguishing a match, for example, tears us away from one group and instantly transports us to another.

1.11 On the Secrecy of Secret Societies

We were saying that the Groupist's hypothesis becomes clearer when we understand which groups we are talking about; that is, groups that are so to speak accidental, that have nothing to do with a decision to belong to them, or the identification that certain individuals may have with a leader or a Sunday activity, a certain type of clothing or a particular cultural belief. But we are immediately disquieted, if we try to get our bearings. Borges and Bioy Casares are basically suggesting to us that every identification works like the kind indicated by the Groupist and that every identification is in a sense accidental. It is never an individual who decides to identify with any trait. If anything, the identification with any trait is what produces the subsequent individuation of that individual. The individual does not identify his or herself, if anything we should say that they are identified, or that they are in the process of identification.

In other words, it is not me who identifies with my dad and chooses as significant that trait of his which is having a moustache, and which in fact I too, decades later, have. Me, who? Is it not precisely a question of explaining the "I", much more than presupposing it? Is the "I" not a product of the identification, much more than a substratum of it? It's always the Westerner who stubbornly opposes the teaching of the Chinese encyclopaedia, always the Westerner who puts being first and the process after, and pretends that he can derive the process from being, with a sleight of hand that takes the place of what should instead be the outrageous courage of magic, deriving of being from process, the assumption that the process is not the proceeding of something but that if anything it is the something that proceeds from the process. The decisive element, in other words, is the emergence of the moustache as a relevant trait, or rather, absolutely relevant. That is to say, relevant not for me who is not yet me, nor for my father who is not yet the father of that "I" that is being constituted, but relevant in itself, relevant without any premise guaranteeing its relevance, relevant on the basis of nothing other than what its very emergence will have made legible in the world, following its emergence.

As we have already said, the secret and the magic of this process entirely consist in the operation of the trait. They consist in its sudden emergence to the eyes of someone who grasps that trait—without being the correlative individual of that trait and without knowing that he or she is the individual who is grasping that trait, but rather becoming it—and at the same time without the individual or the group from which that individual is said to have grasped that trait being for him or her a reference individual or group, since they become one in that same movement.

1.12 The Tactical Nature of Space

A kind of secret plot underlies the Groupist's bizarre list of examples. The first example is based on names. This is the example that suggests a universe with the greatest stability. Catalan names, or names beginning with a certain letter. Membership in this group lasts a lifetime, one would say. But it would be easy to make it crumble down, as Borges did so many times in his stories having a similar subject. What happens if the bearer of a certain name is given a new nickname, after an unexpected encounter with a girl who falls in love with him and renames him, as every lover renames their beloved?

At the other extreme of the Groupist's examples, we find the case of maximum instability, the example of the match. This example is based on a gesture, something that seems destined to be transient by definition. But even this ephemeral gesture could be broken down into smaller and smaller atoms. How do I light the match? By rubbing it from right to left or from left to right? Here are two distinct groups into which I can briefly enter, groups that could of course be further multiplied. At the lowest limit of this series we could identify, using what in mathematics would be called a passage to the limit, the minimum atom of the construction of a group, the simplest unit through which a more or less stable membership in a set is constructed, the irreducible element that sustains a perhaps very short-lived holder of a certain categorisation. Borges and Bioy Casares use a very precise word to indicate this minimal atom, which eludes the idea of substance, the idea of accident, the idea of gesture that already has something too subjective, the idea of event that always risks suggesting something too factual. What they call this minimal unity, it just so happens, is a trait.

1.13 Trait and Threshold

We should think of this trait as a threshold. The trait is a portal through which we pass, like an event that, once immersed in it, makes us no longer the same as we were before we crossed it.

This is what the Chinese encyclopaedia leaves at the margins of our previous reflection, and what we now find at the centre of the story. It is never I who performs a certain gesture; it is never a certain object that runs into a certain event; it is never a given event that crosses a given threshold. It is always the trait that pierces me, it is always the event that awaits another at the opening, it is always the threshold that crosses things, dividing their matter according to new geometries. Put differently, the groups are what passes through me, and not individuals through groups. It is the sets that cut through things by detecting the features that agree with their trait, and leave on the ground the *disjecta membra* that other traits and concatenations will gather in their orbit.

What is absolute, in other words, is the trait through which subjects, things of the world and matter of the universe pass, and what is relative are the subjects, things and matter, which receive a sort of imprint that creates a kinship between them, tactically, as long as they do not pass through other traits, receiving other imprints and entering into other kinships. A whole system of momentary kinships, a whole web of elective affinities is traced with each event. A whole network of negotiations is arranged from each occasion to the next, not only among human beings, as the Groupist would have it, but between humans and animals, animals and plants, humans and minerals and so on, to put it in the words of an encyclopaedia that is now too old, Porphyrian and Aristotelian, certainly inadequate and in need of new formulas. Relative, moreover, does not mean inexistent or inconsistent; it means, if anything, consequent, produced in a kind of immediate filiation, emerging through the movement of an instantaneous phase shift. The trait is one with the quantity of connections it establishes, a quantity that is, so to speak, not extensive since it cannot be compared with something else or something external, but purely intensive, destined to grow or diminish with itself and within itself, and always engaged in stretching out and innervating something other, or rather something that is a kind of non-other of it, a sort of overturned mask of its empty face.

1.14 God, and the Gesture

Let us return to the Chinese encyclopaedia. What governs the universe of this anti-Porphyrian encyclopaedia is not being, and it is not even exactly nothingness. It is the trait, or the gesture of tracing.

What we thus call a gesture, or the gesture of tracing, is that which raises a certain trait to a point of view on animals. What we thus call a gesture, or the gesture of tracing, is that which calls a certain trait to our attention, or to someone else's attention, or to the attention of anything in the universe. Being painted with fine camelhair brushes. Getting madly agitated. Resembling flies from a distance. Lighting a match. Smelling a jasmine flower.

Some might say that something also lies in the background of the emergence of that trait, and that it is not out of nothingness that the category of the animal that gets madly agitated emerges, but from the broader category of animals that just get agitated, or from the even broader category of animals that simply live. Step by step, the Porphyrian desire for consistency comes back into the picture, as does the hope that the tree has a trunk and roots. But is it not clear that even this gradually more general category, this gradually more universal matter, responds to a trait that we have wished to identify, or a characteristic that we have elevated to a point of view over the multitude of beings? If we were to say that all beings have been moulded from clay or have been drawn out of chaos, would not this clay and this chaos also be an image, a deformed mirror, a sort of reversed hypostasis of that gesture of ours in progress, of that action of ours in movement?

That hypostasis, in other words, can be individuated each time because it responds to a certain form of attention. It is the latter's objectivation. Its materialisation, its substantiation. Thus, we can regress towards an increasingly original matter, or rather we can cultivate the illusion of regressing towards an increasingly original matter, which is the same as saying towards an increasingly general space. But it will not in the least be a question of increasingly original matters. If anything, it will be a question of increasingly different matters, because they correspond to traits that are always different. In other words, what the West has conceived as nothingness is an inverted mirror of what it has conceived as god, as we mentioned above. But now we must now go one step further, and say that what the West has conceived as god is what the Chinese encyclopaedia, avoiding substantiating or subjectifying or hypostatising, has conceived as a gesture, as the gesture of tracing, as the tracing structure of every gesture and, to tell the truth, of every human and non-human event. The formula of *ex nihilo* is the symptom, we might say, that within Greek-Jewish-Christian ontology attempts to express—deforming it, displacing it and misrepresenting it—the need to replace ontology with a general theory and a general practice of traits or tracing. *Ex nihilo* indicates, in its own way, a path that goes from ontology to what we might call an ichnology. We might indeed call ichnology an experience of the world as the site of traces, the site of the operativity of traces.

1.15 The Importance of the Letter h

We can now grasp the full importance, which cannot be dismissed as a "well-known logical paradox", in Foucault's words [7, p. XVII] of letter (h) of the Chinese encyclopaedia, that reads "included in the present classification", as well as letter (l), which dryly and comically reads "etcetera".

This is a kind of momentary slap in the face, and that is all it could be. Because it concerns something that cannot be focused on, but only shown marginally, observed as it were out of the corner of one's eye. The Chinese encyclopaedia is not a device that allows us to look out over a certain landscape, with respect to which we would be the spectators. There is no bird's-eye view, no separate point of view. Porphyry's tree is what allows for a bird's-eye view, or deludes us into believing that this is possible, or brings to life the kind of observer who gives the value of reason to that particular madness which from within appears entirely rational.

In the Chinese encyclopaedia "there is no common space", as Foucault said in his own way. He uses also others words. No common "site", no common "ground". This refrain should never be forgotten; on the contrary, it should be put to use on every possible occasion. In the Chinese encyclopaedia, at a closer look, there is one and only one space each time, and that space is always the space generated by a single gesture in action. Looking like flies, or being embalmed, or belonging to the emperor, are each raised to become a decisive trait or a point of view on the universe.

More generally, we could say that each gesture that is at work is a gesture of someone who becomes the great encyclopaedist of the universe. Someone opens their mouth and says: "animals are divided into". And each time, surely enough, in the world, in the universe, there is someone who says something, there is something makes gestures. One way or another, that gesture isolates a certain trait, raising it to a point of observation, making it the cornerstone of its construction of the world, making it a lever on which it can exert his force.

Each time, that someone or something is the centre of the universe; each time, it is the centre from which the universe arranges itself as a universe. Even better, that someone or something is the universe itself, into which other universes plummet and out of which they arise, swarm and dissolve. This means that all animals are indeed "included in the present classification". Also, those who belong to the "other" category, that is, the ones who are excluded because they are exactly complementary to the ones included, and who are therefore included even though they are excluded or precisely because they are excluded. The only thing that actually is excluded, the only great outside allowed by the Chinese encyclopaedia, is the 'we' of the encyclopaedia writing itself. It is the extensionless border that includes the entire extension of the universe, it is the trait that in tracing itself circumscribes everything except itself. Because as soon as you try to say that something is outside that border, that there is something outside the encyclopaedia, that there is something unrelated to the trait that for some reason is establishing itself as a point of view, then you have already included it as excluded, then you have made it the rest of something and therefore the close relative of that something.

1.16 The Empty Empire

What Borges indicates with letter (h) "animals included in the present classification" and letter (k) "et cetera" of his hypothetical encyclopaedia is therefore not a logical paradox. Unless we say that a logical paradox is how the pragmatic, the absolutely contingent and absolutely necessary instant in which it is exercised and the instability of the gesture makes its way into the realm of logic. Ontology yields to ichnology, as we said, and ichnology turns out to be, given the impossibility of tracing an outside without drawing it as an inside, an ethics of immanence.

The trait that occurs, which is singular each time, or if we want to put it this way, the singularity that occurs each time by tracing, has no space in common with other singularities, because it itself is the common space in which every other singularity ends up taking place, ends up occurring or ends up being redrawn, retraced according to the trait imposed by that singularity on it or extracted by that singularity from it, newly collocated and substantiated according to the collocation and according to the concretisation radiated by that singularity. Singularities do not take place in space because they are the place in which a space passes through each time, refolding itself and at the same time unfolding itself. The threshold of every other threshold, the portal through which motionlessly pass other portals to countless other universes.

The imperial city made of the only portal that gives access to it. A universe enclosed in a city that encloses countless others, which in turn enclose it. An intensive empire of traits. An empire which is empty and therefore effective.

But we should correct what we said earlier, that the substances we are talking about are momentary, with no history and no future. True, they are without history and without future, but precisely for this reason they are anything but momentary. They are eternal, and moment or momentariness is the name we are forced to use for this eternity of theirs, into which all time transits each time as theirs, as saved, as the time they will make their own. They may be brief, momentary, less lasting than others, but it is always another momentary and eternal substance that says so, making them its own object, a chapter of its own encyclopaedia, a circumstance of its own event, an extension of its own measure. As for eternity, it only ever lasts as long as it lasts. This banality, this tautology, is its only paradoxically reasonable measurement. Every event lasts one. Not one second or one year or one aeon, which are units of measurement, and therefore relations between something that acts as a unit of measurement and something that is measured against that unit of measurement. No. One, only one, absolutely one. A one that measures every other thing, or that occurs as the measure of every other thing on the same basis, and of course is measured at the same time, on a completely different scene. A completely different scene which is populated by infinite other scenes, in a completely different universe made of infinite universes of which nothing will ever be known by that single universe we are considering. Absolute, amnesic and unaware universe, except that its event will resound with a sort of background noise, an immemorial memory or a blind vision of other traces, also absolute, amnesic and unaware.

References

1. Bible of King James.: Cambridge University Press, Cambridge (2011)
2. Borges, J.-L.: The analytical language of John Wilkins. In: Other Inquisitions 1937–1952 (1952) [English Trans. by Simms R.L.C. University of Texas Press (1966)]
3. Borges, J.L., Bioy Casares, A.: The Groupist. In: Chronicles of Bustos Domecq (1967) [English Transl. by Di Giovanni N.T., Dutton, New York (1976)]
4. Cherchi, P.: "L'enciclopedia cinese" di Borges. In: Storie e linguaggi, no. 4, pp. 103–115 (2018)
5. Descartes, R.: Troisième Méditation. In: Adam & Tannery (eds.) Méditations Métaphysiques; Oeuvres, Tome VII. Éd. Du Cerf, Paris (1904), pp. 48–49 (1641)
6. Eco, U.: L'antiporfirio. In: Sugli specchi e altri saggi. Bompiani, Milano (1990)
7. Foucault, M.: The order of things. In: An archaeology of human sciences (1966) [English Trans. by Gordon, C. Random House, New York (1994)]
8. Frankfurt, H.: Continuous Creation, Ontological Inertia, and the Discontinuity of Time. Cambridge University Press, Cambridge (1999)
9. Givone, S.: Storia del nulla. Laterza, Bari-Rome (1995)
10. Keenan, E.L., Paperno, D.: Handbook of Classifiers in Natural Languages. Springer, New York (2012)
11. Quinzio, S.: Un commento alla Bibbia. Adelphi, Milano (1972–1976)

Chapter 2
FULGURATIO

Silvia Vizzardelli

Abstract The essay develops the theoretical figure of *fulguratio* as a paradigm of discontinuous and non-derivative genesis, one that escapes both causal logic and developmental schemes. The impetus to recover this term from medieval Scholasticism arises from a critical engagement with Konrad Lorenz, who—dissatisfied with the available terms to name the arrival of the new—resorts precisely to *fulguratio*. In dialogue with Nicolai Hartmann, *fulguratio* is thematized as an ontological leap: a founding act that introduces a new level of being without depending linearly on those below, while nonetheless preserving a connection with them. To this theoretical background, Leibniz contributes a grammar of the possible which, though grounded in continuity, nonetheless allows for fissure as a real modality of emergence. All the thinkers here gathered share a theoretical mission: to carve out conceptual space for the logic of the leap, while remaining attuned to the universal rhythm from which that leap takes off.

2.1 Preface

Fulguratio is a term from medieval mysticism that vividly evokes the question of discontinuity in forms of life, the entering-into-existence of something that was not there before. A sudden passage, a lightning bolt indeed, a leap in state and configuration, birthing a novel totality. Starting from this term, we will attempt to rewrite a short trajectory in the history of heterogenesis, the one that from Gottfried W. von Leibniz, through Nicolai Hartmann, goes up to Konrad Lorenz, Georges Bataille, Jacques Derrida, and Jacques Lacan. We will move both backward and forward in time, guided less by chronology than by logic. We start from the observation that contemporaneity is troubled by a lexical search that, far from being a mere issue of terminological choice, denounces a torment of thinking around what remains unthought. In *Die Rückseite des Spiegels* [17], Konrad Lorenz confesses, from the second chapter, that when attempting to account for the process of organic life, one

S. Vizzardelli (✉)
Università di Verona, Verona, Italy
e-mail: silvia.vizzardelli@univr.it

S. Vizzardelli and R. Panattoni (eds.), *Plasticity and Discontinuity*, Lecture Notes in Morphogenesis, https://doi.org/10.1007/978-3-032-03119-8_2

collides with an obstacle: the inadequacy of our lexical heritage. We have many words at our disposal—evolution, unfolding, development, emergence, etc. All of them indicate the unfolding of something already existing although in latent, non-apparent form, as in the case of the cocoon for the flower, or the egg for the hen. As long as we remain in the perspective of ontogenesis, these terms do their job well. But when we try to grasp the discontinuity of something that simply did not exist before, these terms prove inadequate.

The German term *Schöpfung*, often used to denote processes of creation and invention, already in the early Middle Ages bore a double etymology: *schaffen* (corresponding to the Latin *creare,* "to create") and *schöpfen* (to the Latin *haurire*, "to draw from"). Thus, "to create" would mean, in a deeper sense, to draw from an already existing reservoir. A similar fate befalls the term "emergence," frequently employed in contemporary philosophy, which—as Lorenz suggests—evokes the idea of something already preformed suddenly surfacing, like a whale rising from a sea that, just moments before, appeared empty to those observing it in a literally superficial manner.

Fulguratio is a term that dares more, as it alludes to discontinuity—to the leap—even if, while conceding much to heterogenesis, it might initially appear as a magical term, almost suspect in the eyes of philosophical thought and rigorous argumentation. It names a repressed element in thinking, one that, like all repressed content, tends to resurface—often forcefully—precisely when every effort is made to suppress it. The more one is seized by the desire to discern a meaning, to establish connections, to identify anticipations that prepare what is to come or justify it by references to proximate causes, the more the hypothesis of the leap—of generation *ex nihilo*—emerges as the unconscious of thought itself: the other, unruly scene that posits the lacuna as an unsaturable space and as the very matrix of the discrete.

2.2 Invention

Let us begin precisely from the lexical dilemma encountered by Konrad Lorenz in his most ambitious theoretical work, *Behind the Mirror*. After realizing that the available terms used to explain the process of organic life—such as evolution, unfolding, development—were inadequate to account for heterogenesis, Lorenz turned his attention to a term borrowed from medieval theistic philosophy: *fulguratio*, which powerfully evokes the creative act.

Naturally, the challenge was to secularize it, to free it from reference to any higher influence—namely, God himself: "A thunderbolt from Zeus is for the scientist an electric spark like any other, and if we see a spark in an unexpected point in a system, the first thing we think of is a short circuit, a new connection" [17, pp. 29–30]. There is no divine origin in this prepared yet unpredictable leap, and yet the term *fulguratio* situates us unmistakably on the ground of *absolute birth*—a rough, often repressed terrain, but one which Lorenz believes must be explored and probed. Bataille had already led us down this same path, a few years earlier and along different trajectories, when, after visiting the Lascaux cave in the winter of 1953, he sought to convey his sense of awe at the first emergence of art—the dawn of the human species—describing it as *"a sky's lightning, a flash of energy"* [1, p. 22]. Certainly, art would never have appeared on the face of the Earth had it not arisen from the prior, larval phase of *homo faber*; and yet it erupts as an absolute beginning, in an imperceptible instant that cleaves history in two.

It is important to note at once that Lorenz moves delicately between continuity and discontinuity, to the point that his readers and interpreters have engaged in heated debates—some defending the immanence of life, others insisting on a radical discontinuity. In doing so, they overlook the radical dimension of Lorenz's proposal, which can be summarized as follows: the connection that undoubtedly exists between systems is of such a nature that it can withstand both leaps and the interruption. There is, therefore, the possibility of imagining *a link-without-relation*—and this, indeed, is the true stake of his thought. Let us consider one of Lorenz's own examples. Higher organizational systems constitute a functional totality that presupposes the organization of pre-existing subsystems. This is the case in a living organism, whose nature is built upon the arrangement of its constituent parts; it is what occurs between more evolved beings and their ancestral forms; and it is, in principle, what happens between the totality of living beings and the inorganic matter that underlies their existence. Lorenz writes: "In ontological terms one can define the relationship by saying that the whole *is* its parts, and it continues to be so even if, as the result of a series of *fulgurationes* or "creative flashes", it acquires a number of additional system characteristics in the course of its evolution" [17, p. 33].

This formulation demands careful appreciation and a full theoretical understanding. Let us first note the use of the term "fulgurationes", placed alongside the word "additional". The latter (*supplément* in French) is familiar to those acquainted with Derrida's thought, as it closely relates to the concept of *différance*—a topic to which we shall return. Here, Lorenz attempts to lead us onto the slippery terrain of the *link-without-relation.* While it is true that complex systems cannot emerge without support from subsystems, it is equally true that these subsystems, during the process, do not acquire the novel and advanced properties of the complex systems. On the contrary, they tend to simplify and lose some of their features.

Thus, while the totality of the system rests upon the characteristics of all its members, none of the subunits possesses the properties of the whole. The whole is, indeed, a *supplement*, a *fulguratio.* From this, it follows that a detailed knowledge of the system's characteristics does not permit the predication of those that will emerge in a more evolved system. There is a rupture between the two series that no search for a causal link can bridge—for such a search would inevitably lead us into an infinite regress (*recursus ad infinitum*).

What possibility, then, remains? One might be tempted to invoke an irrational or non-rationalizable residue, as does the biologist Max Hartmann. For him, the fact that evolution has produced humans and oak trees in the Old Continent, and eucalyptus trees and kangaroos in Australia, should be classified under the category of chance. One might even be inclined to invoke the miracle: a supernatural and ineffable principle capable of halting our relentless inquiry. This, however, is not

Lorenz's path—he remains, after all, a good naturalist. And yet, we cannot help but notice a certain family resemblance between the term he chooses—*fulguratio*—and the mystical lexicon that traverses the ridge of mystery and ineffability. Perhaps we could replace *miracle* with *prodigy*—a word less inclined to transcend the bounds of nature, yet still oriented toward the unprecedented, and unpredictable. What, then, is at stake? I would suggest the following: it is a matter of allowing ourselves to be seduced by the mystery in order to interrupt the deterministic relation between systems, while at the same time not relinquishing the possibility of a connection between them.

This, for now, is Lorenz's conclusion: "I would prefer to say that no system on a higher level of integration can be deduced from a lower system, however fully one may understand this lower system. We know with certainty that higher systems have arisen from lower ones, absorbing them and containing them like bricks in a building. We also know, with absolute certainty, the earlier stages in development from which higher living beings emerged. But each step forward has consisted of a *fulguratio*, a historically unique event in phylogeny which has always had a chance quality about it—the quality, one might say, of something invented" (ibid., p. 35).

Between systems, between the layers of being, there exists an unsaturable discontinuity: the same discontinuity that emerges within cognitive processes. A radical reformulation of what we mean by *knowledge* is therefore required. For now, it is enough to note that *fulguratio* is nestled precisely in the moment when we feel we are grasping something—when transcendence is entirely internal to the act of knowing. We must fully appreciate the originality of this perspective. Traditionally, we have been accustomed to think of knowledge in two ways: as a *bridge* between subject and object and as an *inevitable categorical distortion* of the object.

The subject reaches toward the object, striving to close the gap between them, yet in doing so, it grasps only what appears, only what it is capable of perceiving. According to this paradigm, to know is to pursue the object, only to witness it recede. This is the Kantian paradigm, so prominent in Lorenz's thought, and one that ought to be preserved provided it is reoriented toward an entirely different meaning. To read Kant properly, according to this new approach, is to sequentially abandon all the *idola* of reason that have burdened interpretations of criticism: the idea that knowledge seeks continuity between subject and object; that this search fails because consciousness remains imprisoned within its own forms, and that the object remains forever inaccessible, retreating into the thing-in-itself (*Ding an sich*).

What alternative can be posed to this paradigm? A formulation nearly identical in wording, in enunciation, but completely different in spirit, in its point of enunciation. It is true: knowledge is an act reaching toward what eludes it. And yet—and this is the crucial point—this aporia does not negate the act of knowing; it constitutes its very foundation. The task is not to delineate the limits of knowledge, but to recognize it precisely at the moment of its apparent failure: we sense that we grasp something of the real precisely at the point of our *mis-taking* (*méprise*). The insoluble, the discrete, the singular, the unanticipable, and the transcendent arise not as negations, but as affirmative figures of knowledge—anything but mere placeholders of negative theology.

2.3 Discontinuity

At this point, the biologist, physician, and ethologist Konrad Lorenz introduces one of the most intriguing philosophers of the twentieth century: Nicolai Hartmann. Trained in the neo-Kantianism of Marburg, Hartmann succeeded in redirecting this tradition toward an innovative ontological approach. This reference is surprising for at least two reasons: Hartmann remained in the shadows for a long time, often unfairly regarded as philosopher who took a step backward in the history of thought—a proponent of a naive, pre-critical realism; despite his engagement with various fields (ethics, aesthetics, philosophy of history), Hartmann's central contribution lies in ontology—a domain that might appear distant from Lorenz's scientific concerns.

How, then, can a scientist draw inspiration from this philosophical universe? The fact that he does suggests a certain measured optimism regarding the contemporary debate between philosophy and science. Clearly, Lorenz sensed that he shared what Deleuze might call a "common cause" with Hartmann. And it is precisely this common cause that allows for a dialogue between a phylogenetic and an ontological perspective. Lorenz's interest in Hartmann thus shifts from epistemology to ontology, recognizing an indissoluble link between the two. In both cases, the challenge lies in accepting the paradox of a discontinuity (whether between subject and object in epistemology or between layers of being in ontology) that does not preclude relationship. Phenomenological analysis characterizes knowledge as a direct, frontal relation between subject and object. This relation is knowledge itself, yet it never dissolves the oppositional structure (*Gegenüberstehen*) that preserves their mutual separateness—a fundamental transcendence within the relationship itself. "The *Gegenüberstehen* thus sustains both the inescapable and essential relationship between the knower and the known, and, at the same time, their original separateness—the transcendence of their being-in-relation" [2, pp. 32–33].

Similarly, Hartmann's ontology recognizes the presence of distinct layers of being (the inorganic, the organic, the psychic, the spiritual)—each maintaining a connection without merging into a single continuum. Lorenz focuses on Hartmann's foundational work, *Der Aufbau der realen Welt* [8], to highlight the analogy between Hartmann's ontological stratification and the phylogenetic theory of *fulguratio*.

In Hartmann's thought, the layers of being are independent, autonomous, and discontinuous, yet they are linked through a one-way, asymmetrical connection. Organic nature rises above inorganic nature, emerging from it, with the relationships and laws of inorganic matter as its necessary precondition, though not sufficient to give rise to the living world. Similarly, consciousness and psychic life could not exist without their connection to the organism that embodies them; yet the physiological characteristics of the organism in no way anticipate those of psychic reality. The relationship between layers is thus one of support and conditioning from below, while preserving the full autonomy of the emergent layer. Individuality, autonomy, and independence do not exclude connection and relationality.

This paradox is what fascinates Lorenz and underlies his enthusiastic reception of Hartmann's ideas. What is striking is that the notion of a *link-without-relationship*, of a continuity that coexists with an absolute beginning, appears at first glance to be an abstract formulation—the result of pure speculation rather than empirical discovery. Yet Lorenz argues that this insight is not a logical construct, but an empirical and phenomenological reality. Experience itself reveals both the absolute heterogeneity of life's events and their linkage across discontinuities. It is experience—not theory—that reveals the indeducibility and unpredictability of emergent phenomena. Lorenz is struck by the ontological precision of Hartmann's theory and its remarkable alignment with evolutionary biology, even though Hartmann never explicitly addressed evolutionary thought. Each advance in phylogenesis shares the same nature as the original emergence of life itself: an invention, a flash that relies on the preceding level without ever being anticipated or deduced by it. Throughout the history of thought, many "isms" have sought to distill the complexity of reality into a singular principle—whether by gazing downward (mechanicism, biologism, materialism) or upward (spiritualism, panpsychism, idealism). Lorenz argues that both reductionist and holistic paradigms ultimately falter insofar as they neglect the stratified nature of reality. The mistake consists in "trying to explain the diversity of the world on the basis of ontological or phenomenal principles of one single kind" [17, p. 40]. Drawing upon the ontological insights of Hartmann, Lorenz arrives at a pivotal recognition: both the compulsion to unify reality under a monolithic framework, and the inclination to dissociate its layers entirely, represent epistemic missteps. The philosophical task, then, is to dwell within the tension, avoiding both reductive unity and absolute fragmentation. In this spirit, Lorenz articulates two sets of statements, apparently antithetical yet jointly indispensable:

1. A reductionist-deterministic claim:
   - "All life processes are physico-chemical phenomena."
   - "All subjective experiences are organic, physiological, and thus physico-chemical."
   - "All spiritual life unfolds within the biological and physical substrates that sustain it."
2. An emergentist claim:
   - "Life processes are something completely different from physical and chemical processes."
   - "Consciousness is completely different from non-conscious neural activity."
   - "The human being, as a cultural and self-aware entity, is fundamentally distinct from its closest zoological relatives."

These two sets of statements are not truly contradictory: they run in parallel and both express valid perspectives. By rejecting false dichotomies, Lorenz's approach undermines many of the polarized and sterile debates between scientists and humanists, which have long failed to yield meaningful epistemic progress. Lorenz concludes:

> The seeming contradiction of these two sets of perfectly true statements indicates the existence of a pseudo-problem, which can prove a serious hindrance to the advance of human

> understanding. Its resolution is one of the most important results at which, independently of each other, the ontology of Nicolai Hartmann and the causal-analytical investigation on living systems have arrived: namely, that the law of contradictory opposites does not apply to the unilateral penetration of strata of levels of integration. *B* is never non-*A* but always *A*+*B*; *C* is *A*+*B*+*C*, etc. Although it is inappropriate to use mutually exclusive concepts when dealing with different levels of existence, numerous such pairs of terms have crept into our thinking and are to be found in both our scientific and our everyday language—mind and nature, body and soul, man and animal etc. (ibid., pp. 41–42).

We could thus say, in summing up this discussion, that *fulguratio*—precisely as a flash of discontinuity grounded in antecedents, which, as repeatedly noted, are utterly incapable of prefiguring even the slightest characteristics of what follows—constitutes the very locus of thought we are seeking, precisely because it renders the two series compossible. Continuity and discontinuity are not two external properties, as if there existed a universal, an essence, or a unity poised to individuate itself, to differentiate, and only later to give rise to distinct entities or ontological layers. The universal is already individual, and the individual is already universal. The layers of being are unique and unrepeatable precisely because they preserve a relation among themselves. The task, then, is to grasp the nature of this relation.

To delve into the substance of this dizzying proposal—which has reshaped our way of thinking by offering an escape route from the binary of identity and opposition—it is helpful to focus on a very specific and largely overlooked episode in the history of thought—with a few fortunate exceptions (see [7]. We must rewind the tape of time to July 2, 1946 [9], when Nicolai Hartmann delivered his solemn address during the celebrations for the three-hundredth anniversary of Leibniz's birth, organized by the University of Göttingen, where Hartmann held a professorship. Strictly speaking, this retrospective glance at Leibniz, is not endorsed by Lorenz, who, oversimplifying and misjudging, groups the philosopher from Leipzig with Jacob von Uexküll and W. Weidel among those who, in his view, commit the error of attempting "to explain the diversity of the world on the basis of ontological or phenomenal principles of one single kind" [17, p. 40]. It is therefore preferable to follow Hartmann's lead.

2.4 Fissure

In his solemn address delivered on July 2, 1946, Nicolai Hartmann opens by highlighting what fascinates him most in Leibniz's thought: its extraordinary aporetic vocation—a refusal to be guided by a pre-formed, systematic worldview, and a willingness to let the investigation itself lead the way, embracing contradictions and antinomies rather than prematurely resolving them. There is a fissure (*Riß*) that runs through Leibniz's philosophy, and Hartmann is drawn to his capacity to endure this rupture—without rushing to mend it, without attempting to resolve it through a pacifying third synthesis. Most interpreters have taken this aporetic structure as a sign of theoretical indecision. Hartmann, however, reads it as a mark of vitality, originality, and visionary depth—a capacity to hold together all the competing positions in the historical dispute over universals, without neglecting either side.

At the core of Leibniz's unresolved tension lies the contrast between a metaphysics of universals and the individualism of monads; between continuity and discontinuity; between the essential components of reality (*requisita*), its foundational and simple elements (*simplices*), and its individual ontological unities; between the *veritates aeternae* and the indivisible monads. Leibniz does not seek a reconciliatory synthesis between the universal and the individual. Rather, he remains within the oscillation between them—that minimal difference which both separates and binds them. Hartmann grasps this point with great precision. He sees in Leibniz's work something deeply resonant with his own ontological project. Leibniz's innovative synthesis is not the post hoc reconciliation of two originally distinct perspectives; it arises because universalism, when pursued to its ultimate consequences, does not culminate in a merely numerical, quantitative individualism—but instead in a "pure realism of universals". Here, the universal manifests itself as individual. This is a *pure* realism because universals do not constitute an autonomous sphere existing alongside individuals. Unlike classical realism, here the individual is not a mere instantiation of a pre-existing universal essence. The monad is, rather, the very pulse of the universal itself. It must be emphasized that the universal at stake here bears little resemblance to that of traditional metaphysics, which rarely escapes an axiological framing. It is instead the name of a living process that passes through individual bodies, or better yet, leaps from one to other, always reappearing, identical and yet embodied anew. For both Leibniz and Hartmann, the ontological unity of the world is not a homogenization of differences, but the very principle of radical and irreducible dissociation—the uniqueness of monads (Leibniz) and of ontological strata (Hartmann). One could thus say that the identical is a singular, recurring moment that appears in all entities, yet, in each, it is unique and unrepeatable. It becomes clear, then, how the discourse of continuity can intersect with that of discontinuity, making *aporia* the most precise, and ontologically adequate, instrument of thought.

Returning to Lorenz's interpretation of Hartmann, particularly concerning the relationship between human and animal, we must acknowledge that the human animal and the non-human animal are both unique and unrepeatable. Between them lies an unsaturable *laceration*, an absolute discontinuity—the *fulguratio* that illuminates our inquiry.

It is deeply misleading to attempt to characterize human nature by reference to animal nature, or vice versa. And yet, in both, the same pulse of animality resounds. Animality is a universal trait, but it is not a universality that rests in an autonomous, self-sufficient sphere coexisting alongside its instances, as real universalism would claim. Rather, it manifests only and always through individuality. Every being is defined by its unrepeatable singularity, bearing within it the aura of the universal—a resonance that persists in the very act of giving voice to difference. To illustrate this, consider the following two statements:

a. Life that lives manifests as individuality;
b. Life is individual.

The first formulation differs radically from the second. In the first, there is a doubling: while we witness the *fulguratio* of individuality, we also sense the halo of life itself—the echo of its source. Every form, every figuration, carries the aura of its birth, the moment of its coming into life. Perhaps it is better to remain silent about origin. How much harm has been done in attempting to ground the human, or the living, in some original blessing or curse? How much damage has followed from the belief that, somewhere in the flux of life, there exists a privileged point—a fixed origin dictating the course of all that follows?

And yet, *origin* is a beautiful word. It tells us that we are born from where we are not, that life is not merely itself but carries a shadow, a blurred point rendered so by motion. The living being is a projection, a continuous displacement, a perpetual deferral. We are nothing other than what we become, moment by moment, and yet every form, every figuration, still carries the aura of its birth, of its coming into life. And it is good that this aura makes itself felt, seen, spoken, that it produces effects. An aura is a quasi-nothing; and yet how, then, does it appear? How does it take on form? The fertile mystery lies entirely in that "quasi": in it, the possibility of sensing the pulse of the universal is at stake—without letting it dissolve into the labyrinth of the continuous. What is at stake is a shadowy consistency, a little more than zero, a little less than one, that accompanies every being, whispering that it never fully coincides with what it is, moment by moment. In the history of thought, we have sought to name this aura: *world-sentiment, the transcendental, the possible, the virtual.* Each of these, in different ways, reminds us that the contingency we are and embody is not merely contingent. Its aura lends it a kind of necessity. What matters is not to grant that origin an autonomous substance. It is best that it remains an indication rather than a substance, a scent rather than a sentiment, a resonance rather than a nostalgia. We might call this aura a "secret", as Derrida did. Not everything is exposed in the light of the manifest, but what withdraws from it is not elsewhere, rather, it insists and pulses within the very event of appearance. Derrida writes: "I can't quite recall at which moments, but there are moments when I walk, I can't be seen speaking but my voice can be heard. There is a speech more ancient than me, older than that moment, that carries on speaking, that precedes me, and it is this body that is running after its speech or speech that is running after this body that indeed gives the feeling of a secret, of the interruption that forbids (*interdit*), that in short speaks in between (*entredit*)" [3, p. 66].

The aura, the secret, is akin to what acoustics refers to as the attack transient (*transitore*): a very rapid amplitude variation occurring at the onset of a sound, before it reaches any form of stability. The attack transient of a real-world sound—like a played note, a drum hit, or an explosive consonant—is an event that is fast, intense, but not instantaneous: it has a finite duration, and often a complex spectral evolution. However, in many simplified or abstract models, the attack is often approximated as an instantaneous impulse in time and interpreted as a trigger that excites a system—like a hit that sets a string or membrane into vibration. The hypothesis I'd like to develop here is to give credence to this symbolic and abstract model, because it brings us closer to the meaning of heterogenesis. Strictly speaking, there is not an intermediate point that fills a gap or acts as a bridge, as if there were something between silence and sound that could mediate their passage. Rather, it is an *attaccostacco* (attack-detachment)—a neologism, a single word that expresses both sound and its silence. It is a sound that lets us hear the silence from which it emerges, and a silence that lets us hear the imminence of sound. This is why we often speak of the breath of sound. We might put it this way: *at the inception of sound waves, there is an explosive burst of energy that marks the wave's initial peak. Its explosive, triggering nature makes the transient that very aura of the "from where"—the hidden origin that constitutes sound's secret.*

In thus the *Dirac delta function* offers more than a mathematical analogy. It stands as the paradigmatic symbol of such a phenomenon. The Dirac delta is a mathematical distribution: it is zero everywhere except at a single point, where it is infinitely high, yet its total integral is one. It is an excellent model for describing a pure generative event: it does not exist as duration, but it happens, and its effectiveness is real, measurable, and full. In other words, it models a perfectly instantaneous impulse, a pure event without duration, but with maximal intensity and generative power. It does not exist in time, but triggers all that follows in time. The delta is not a state but a singularity, a punctual infinitely concentrated event that nonetheless produces effect: it is the formal figure of the *attaccostacco*, the gesture that opens the field of the audible. This type of approach is mainly found in more formal and theoretical contexts of physics, signal theory, and computational music, where the emphasis is placed on a mathematically ideal representation of sound. Let us, now, imagine applying a very slight modulation to the Dirac delta, in such a way as to let us better perceive the point of attack. I am convinced that this approach may be much closer to the realistic perception and production of sound than one might be led to believe. Something very similar was suggested by Iannis Xenakis, who—thanks to the strength of an initially intuitive representation—did not conceive of sound as a continuous flow, but as a multiplicity of discrete events (see [21]).

But let us pause on this point: if we can perceive the universal breath of the being (sound) only at its attack point, this means that there can be no *continuity* between beings (extremes that are a single extreme, *en*), but only *contiguity* (two extremes simultaneously, *ama*). Or, put another way: the first thing we perceive about a sound is at once the end of silence and the beginning of sound—not something that exists between silence and sound. In phonetics, this concept is perhaps more evident and immediately graspable. The *voice onset time* (VOT), which characterizes unaspirated stop consonants, is the time that passes between:

a. The release of the consonant with articulatory closure (that is, the complete blockage of airflow at the level of the mouth, pharynx, or glottis), and
b. Its sonorization, meaning the rapid release of the blockage.

For this sound to occur, it is necessary to close and open. The quality of the emitted sound simultaneously reveals both: the end of silence (articulatory closure), and the beginning of sonorization (release and vibration of the vocal folds). In a voiced consonant, we perceive *at the same time* the attack and the detachment (*attaccostacco*).

It is the atomic and nominal nature of the sound that reaches maximum amplitude, establishing itself as a singularity, that retains its virtuality, its transitory attack. Affirming this, however, means embracing another type of narrative, or rather changing its point of balance. The attack transient is generally described this way: any phenomenon in which a certain amount of energy is involved cannot abruptly pass from one energy state to another. Thus, the sound cannot abruptly go from silence to maximum amplitude. This is a description that evidently relies on the "before" (the transitory) to explain the emergence of the "after" (the nominal sound). What is at stake, however, following our discussion, is to shift the emphasis to the singularity of the sound at its maximum amplitude in order to hear its point of attack. It is the point of attack that guarantees the minimum difference between being and entity, because it is not the difference between two terms, but the generation of a difference through an act. The transient, like the secret or the aura, does not mediate: it *initiates*. The aura that precedes the maximum amplitude of the sound is nothing other than its very point of attack. And it is this point that gives density to the sound's onset. It does not resolve itself into the punctual entity it is, because we hear its transition, its exercise, its pure sonic potential. It is the ontological switch, the inaudible force that gives rise to the audible. Here, Bruno Latour's ontology becomes illuminating: being is not what lies beneath or before, but what circulates, *passes*, emerges through actions and insists through effects. The attack is not a phase in a process, but the activation of a network. In this sense, the delta function is not merely a metaphor, but a formal expression of a Latourian actant, an operative unit of change, whose very non-being is what makes being possible (see [12, 13, 18]).

Beyond this example which, like all examples, could be replaced by others, perhaps better ones—what lies at the core is an issue particularly dear to the young Leibniz, one that was, even then, regarded as both insolent and aporetic, and ultimately abandoned, though not without significant reconsideration.

We cannot overlook Leibniz's defense of continuity, condensed in the expression *"tout va par degrés dans la nature et rien par saut"* (*Nouveaux essais sur l'entendement humain*, IV, 16, §12), a translation of the Latin *natura non facit saltus*. Yet in a 1676 dialogue, *Pacidius Philalethi*, Leibniz confronts a persistent difficulty: if transitions in nature mirror the continuous transitions in geometry, how can we move from geometry to physics, where, conversely, one must account not only for continuity but also for the discontinuities inherent in physical change. Leibniz is thus in search of a science of motion that can hold together the speculative nature of continuity with the practical discontinuity of physical change. In other words, physical transformations resist complete explanation through the logic of continuity. As Larry M. Jorgensen emphasizes, Leibniz arrives *reluctantly* at the following three conclusions:

a. The link between geometry and physics must be via mind (whether human or divine);
b. In the absence of mind there would be radical discontinuities in nature;
c. Given the intervention of mind into the material world, some (less radical) discontinuities remain [10, p. 74].

Despite Leibniz's efforts to counter atomistic models of motion—those built on discrete, indivisible units—the leap, the discontinuous, symptomatically resurfaces. This is why, at the beginning of my text, I wanted to emphasize how discontinuity emerges from the unconscious of thought. The hypothesis of the *leap* emerges clearly in *On Motion and Matter*, a short essay from April 1676. It proposes that motion entails not just transition but actual re-creation: the moving body must, in passing from one location to another, be destroyed in one place and reconstituted in another. It is a necessary absurdity required to circumvent the aporias of a strictly continuous model, which cannot explain phenomena like relative velocities without presupposing the division of a line into infinitesimal points. In a word, if motion were continuous, *things would produce themselves.* At this point, Leibniz introduces the concept of *transcreation*, being forced to admit that a body in motion is transferred from one place to another without passing through an intermediate space, meaning that matter ceases to exist here and reproduces itself elsewhere. It is a hypothesis that borders on the miraculous—a kind of vital magic, wherein the living dies in one place to be reborn elsewhere. Leibniz would soon renounce the idea, but it nevertheless offers a vivid lens through which to grasp the non-synthetic entanglement of universal and individual, continuity and discontinuity. Faced with the threat of fragmentation posed by discontinuity, Leibniz introduces the idea of a divine mind as the intact cause of all things—not one that creates once, but one that continually recreates, incarnating itself within the rhythms of nature. Jorgensen sees in this move a proximity to occasionalism, the doctrine that attributes all causality to a universal divine agent. I would suggest, however, that something different is emerging: it is not a matter of viewing the world as an incoherent succession of leaps and entrusting the guarantee of continuity to a divine mind, detached in an autonomous sphere and guardian of the universal. Nor is it about hypothesizing an occasional dependency between phenomena and a single cause represented by the action of the universal divine principle. Rather, it is about feeling God—or, more secularly, the universal—*as the aura of the individual.* In this light, we must regard as compossible (recalling Lorenz's two compossible series) both the series governed by the leap and that governed by *non facit saltus.* That is: matter is not a passive, inert, neutral substrate awaiting form, but the very pulse of the universal within the individual. In the dialogue *Pacidius Philalethi*, written between October and November 1676, while Leibniz was stalled for six days in the Thames estuary, waiting for good weather to complete the sea voyage that would take him from London to Holland, the philosopher questions the *Metaphysics of Motion.* The dialogue unfolds between *Pacidius*, an alter ego of Leibniz himself in the Socratic role of interrogator, and *Charinus*, a promising student reflecting on how to establish a relationship between geometry and physics concerning the issue of motion. If geometry describes motion in a continuist key, seeing *circles* where physics instead sees *polygons*, the problem remains that of making the formal and the material speak to each other.[1] It is not,

[1] "Charinus: If I may be allowed to offer an inexpert opinion on such matters, I would declare that the transition from Geometry to Physics is difficult, and that we need a science of motion that would connect matter to forms and speculation to practice—something I learned from experiments of

therefore, a matter of choosing between one or the other, but of conceiving of a mind—in this case, the divine mind—that holds together the *continuous* and the *discontinuous* as two compossible dimensions. To achieve this, one must invoke the *contiguous*. The concept of *transcreation* allows Leibniz to identify *universal intelligence* as the support for *motion by leaps*. The activity of *God's transcreation* causes a body to be *destroyed in one place* and *reborn in another*. This means that it is not enough to describe this passage generically as a *leap*. We know, in fact, that we usually think of a leap as a *skipping over intermediate intervals*. What is at stake here, however, is a very particular leap—one that does *not pass through intermediate places*, a leap that may seem like a miracle when macroscopic distances are involved, but that becomes more acceptable at the microscopic level.

Leibniz is here attempting to distinguish the concept of *continuous* from that of *contiguous*, drawing on Aristotle's distinction (voiced by *Charinus*): *continuous things share one and the same extremity, while contiguous ones are simply together at their limits*. From this, follows, for example, that being alive and being dead are contiguous, because they have no common extremity; their lack of a shared extremity creates a leap, a dense instant in which both states coexist. How are we to imagine such a moment? That is the central challenge Leibniz confronts. The lack of anything in common, rather than excluding, tolerates and founds the possibility of coexistence, of simultaneity.

It is true that Leibniz very soon abandons the term *transcreation*, delving instead into what he would later call the *labyrinth of the continuous*. When he briefly revisits it in 1709, regarding the *generation of rational beings*, it is to describe God's addition of reason to a sentient soul, not to endorse a discontinuity in the chain of being.[2] And yet, I would argue that a residue of this *theory of the leap* persists. Even in his most committed phase of continuism, the trace of discontinuity remains. Rooted in an intellectual climate still shaped by Aristotelian theories of spontaneous generation, only beginning to be displaced by discoveries such as Leeuwenhoek's spermatozoa in 1667—this trace reveals something crucial. However, we must accept that *the descent into the microscopic*, rather than revealing the *changing movement of the continuous*, leads us to a place *almost unthinkable* and yet much pursued by contemporary thought, where the continuous spasmodically presents itself as the pulse of the discontinuous.

---

various kinds in my early military training. For I was often unsuccessful in trying out new machines and other delightful tricks of the trade, because the motions and forces involved could not be drawn and subjected to the imagination in the same way as figures and bodies could" (leibnizsdialogue.pdf).

[2] See Jorgensen Larry [10]: "With this background in mind, it is surprising to find Leibniz returning to the concept of transcreation. Elsewhere I have argued that Leibniz does not *need* to appeal to transcreation and that, when he did, he always left himself a way out. And so, ultimately, I am doubtful that transcreation ever played an important role in Leibniz's philosophy after 1676. However, it is nevertheless interesting that Leibniz appeared to think that transcreation in limited cases was at least *consistent* with his mature philosophy" (p. 87).

2.5 The Lightning of Lascaux

Let us shift the scene and draw closer to our own time. The story is well known. It all begins on September 12, 1940, when four boys, aged between thirteen and eighteen, descend into a narrow tunnel—a magical, devouring mouth, opened by the uprooting of a tree after a storm—in search of a small dog named Robot, who had disappeared beneath the juniper bushes.

We are in Lascaux, in the Dordogne region of southwestern France. Suddenly, within the darkness of the cave, they discover a miracle: drawings and engravings emerge on the walls—over six hundred paintings, masterpieces of Cro-Magnon humans. What would later be called the "Sistine Chapel of the Paleolithic" depicts oxen, bears, bison, horses, deer, a unicorn, and a single human figure: a bird-headed shaman confined in what is known as the "well scene.“ Before the cave was closed to the public on April 18, 1963, by the then Minister of Culture André Malraux, who was concerned about the marks left on the fragile rock walls by the carbon dioxide exhaled by tourists, it was visited in the winter of 1953 by an illustrious emitter of carbon dioxide: Georges Bataille. That was a fascinating expedition, one that still evokes nostalgia for a time when research was a strange mixture of seriousness, adventure, and friendship. Alongside Bataille, into that cave entered the Genevan publisher *Albert Skira*, and the Swiss photographer Hans Hinz. The photographs, taken in a state of wonder, and the Bataille's commentary were published in a magnificent edition by *Skira* himself (*Lascaux ou la naissance de l'art*).

Bataille describes the images as a miracle, a prodigy. And it is precisely this notion that concerns us: something is given like lightning, and philosophical thought is truly such if it dares to incorporate it into its discourse—perhaps by secularizing it, making it compatible with a naturalistic perspective, as did Leibniz, Hartmann, and Lorenz. The miracle described by Bataille is what allowed the human being to distinguish itself from the beast: the miracle of anthropogenesis. A prodigy, a leap, a flash of lightning—a moment of absolute discontinuity that enabled the birth of the human being through art and play. But how is such an absolute and discontinuous birth represented? One might expect the newly arrived human to dominate the scene, to demand all attention.

And yet, among the over six hundred paintings in the cave, only one human figure appears, lying on the ground, possibly dead or wounded, while all the rest foreground the overwhelming, luminous presence of animality. Bataille does not evade this aporia, just as Leibniz did not, according to Hartmann's penetrating interpretation. They are thinkers who endure contradiction, who remain within the tension of paradox rather than feeling towards synthetic resolution, because it is within aporia that truth is at stake. Bataille writes:

> These Lascaux Men forcefully transmitted to us the fact that, being men, they resembled us, but as a means for telling us so they left us innumerable pictures of the animality they were shedding—as though they had felt obliged to clothe a nascent marvel with the animal grace they had lost. These non-human figures, wrought with youthful strength, declare not only that they who painted them became full-grown men by painting them, but that they

chose animality rather than themselves to give the image that suggests what is fascinating in humankind [1, p. 119].

Let us try to interpret this in light of what we have said so far. There is a universal (*animality*) that pulses alongside singularity (the human animal, the non-human animal). The continuity—here highlighted by the need to depict the past (the imposing and radiant bestiality) precisely at the moment when the new emerges (*Homo sapiens*)—does not negate the *fulguratio* of spontaneous generation. Quite the opposite. Hence the setting: a cave. A topological space where darkness coexists with the intermittent light of lanterns, life with death, inside with outside, animality with humanity. We must resist interpreting the cave through a Romantic or Jungian tropes (as an archetypal depth) or Platonic allegory (as the domain of illusion and shadow). These dualisms must be surpassed in favor of a topological articulation, one that, as Lacan suggests—with explicit reference to Paleolithic caves—sees the cave not as a vault of truth but as something more organic, more paradoxical: a bladder, which on the condition that one places a little light inside, can serve as a lantern. "It is usual", says Lacan, "to represent the unconscious as a cellar, even as a cave, by way of allusion to Plato's cave. But it is not a good comparison" [11, p. 187]. The bladder is better. It is this that allows us to hold together, in the same instant—like two beats of the universal—inside and outside, darkness and light, life and death (see [16, pp. 114–137].

2.6 Bladder

Bladders, lamellae, membranes aptly describe the tight, spasmodic oscillation of opposites. When Derrida, for example, attempts to account for the simultaneity of *inside-outside, within-without, present-absent*, he turns to Mallarmé, inheriting from him the spasm of the hymen. In Mallarme's text *Mimique*, published without a title in 1886 in *La Revue Indépendante* and inspired by Paul Margueritte's pantomime, *Pierrot assassin de sa femme* (1882), the term "hymen" refers to the mime's gesture. It is that membrane, both impure and sacred, that inserts itself between *desire and fulfillment, between perpetuation and its recollection.* It is important here to underline the oscillation, the pulsation *sur place* between presence and absence that defines the mime's action. There is no entity to be imitated, no transcendent form or original presence—if one follows a Platonic or metaphysical understanding of mimesis. No model governs imitation; no image conforms to a prior truth. Derrida characterizes Mallarmé's departure from this classical paradigm as a subtle, patient, and efficient displacement. The mime does not reveal a presence, nor does it represent a hidden truth. Rather, the mime illuminates the white space, foregrounding the interval, the spacing from which the gesture itself arises. "*Hymen*" is the term that comes to name this threshold, a "supreme spasm" between absence and presence. But what is at stake in this spasm? Derrida approaches this restless locus, a site where the boundaries of classical logic falter—in his text *La double séance*, first published in *Tel Quel*

(1970) and later included in *La dissémination* (1972) [3]. It is the place of a marriage consummated, yet never completed; a union that does not resolve into unity. The hymen first marks fusion, the identification of the two into one: the disappearance of the gap between desire and fulfillment, expectation and presence, longing and consummation. The hymen simultaneously opens and closes. This simultaneity is decisive in describing the spasm between presence and absence. "*Nonpresence, the gaping void of desire, and presence, the fullness of enjoyment, amount to the same.*" [3, p. 209]. Likewise, in the mime's gesture, there is no longer any textual difference between image and thing, imitator and imitated. Yet it is not enough to emphasize this fusion. The risk is that the "*hymen of confusion*" might lead us into the realm of sameness. One might legitimately propose the misleading hypothesis that presence and absence, the imitator and the imitated are no longer two, but have merged into a single term, one we might then feel compelled to uncover or invent. Derrida, however, observes, with great acuity, that confusion does not imply unity. Difference is not erased. What is erased is the decidable heterogeneity of the two realms, independent and irreversible. "*Thanks to the confusion and continuity of the hymen, and not in spite of it, a (pure and impure) difference inscribes itself without any decidable poles, without any independent, irreversible terms.*" (ibid., p. 210).

Between the autumn of 1975 and June 1976, Derrida held a seminar at the École normale supérieure entitled *La vie la mort*. This seminar formed part of a training program designed to prepare students for the agrégation in philosophy. As such, Derrida was therefore not entirely free to determine the content or theme of his teaching. For the 1976 examination, the prescribed topic was "*La vie et la mort*", but Derrida immediately announced that his intention would be to offer a counter-seminar, one that would interrogate and deconstruct the very title itself. At first glance, his intervention might appear minimal, a mere semantic shift, since he simply proposes to elude the *et* in the title, transforming it into "*La vie la mort*". Yet this elision marks the inauguration of a deconstructive gesture aimed at the oppositional logic that underpins classical philosophy. Derrida invites us to resist conceiving life and death as opposites; rather, we are to understand them as *differents*. The absence of punctuation between the two terms, and the bare insertion of spacing, acts as an invitation to exit the logic of identity and opposition—the logic of *et* and *est*, which animates Hegelian philosophy in its movement from contradiction to the self-same synthesis. What Derrida gestures toward is the possibility of a new mode of questioning, one that deconstruction aspires to: a mode that maintains its distance from both monism and dualism (mind–body, life-death). The paradoxical force of this formulation is immediately apparent, producing an unsettling effect. That life is swollen with death means, first of all, that death does not stand as life's dialectical contrary. But neither is death identical with life. To even begin to grasp this "other logic", one must think ventriloquially—a form of thought that, in speaking, makes silence felt; that alters itself mutely as it thinks; that, while speaking, experiencing pleasure and living, yet at the same time, falls silent, suffers, and dies. Perhaps here we begin to trace the philosophical function of membranes—and not only within philosophy. Psychoanalysis, too, has long attended to these minimal thresholds, these fragile tissues shaken by a logical spasm. Here, then, is another example, beyond that of the bladder.

When Lacan evokes the trauma of birth and the anguish that descends from the release of libido, he employs a curious image. Still within the frame of *Seminar XI*, devoted to the four fundamental concepts of psychoanalysis, he offers a striking wordplay: in order to make the *Hommelette*—a portmanteau of *homme* (man) and *homelette*—the egg must be broken. In that rupture, something is released: the *lamella,* which begins to circulate on its own—an uncanny embodiment of libido [11, p. 197]. Even here, it is crucial to emphasize that a spasm is produced between pleasure and displeasure, between absence and presence, between void and fullness. These are oppositions that the principle of non-contradiction warns us not to allow to coexist. Yet they do, precisely in the zone Lacan opens up. The immortality of the lamella, its status as an acephalic representative of libido, a nonhuman tissue, can appear within the finite and mortal horizon of the human only as a formation that saturates and substitutes for the ontological void left open by the finite subject. There is something immortal, a pre-subjective, pre-sexual substance, irreducible to the digital order of the symbolic, which, however, in manifesting itself in the world of the finite, must assume some kind of organ structure, even if it is a paradoxical organ, evoking "a life that needs no organs".

The lamella, like all membranes, all minimal tissues, all thresholds, vibrates with a spasmodic resonance, in which immortality and finitude, absence and presence, coexist in a kind of obversion. The lamella is a found object, already lost in itself, *thing/non-thing*, *thing of jouissance*. The logic of mutual exclusion is replaced by the logic of contradictory inclusion—not a reconciliatory dialectic, but an ontologically incestuous tension. But why do Lacan's bladders and lamellae, or Derrida's hymen, prove more operative in the economy of the discourse we are attempting, than the Platonic cave? Precisely because the cave remains too structural. It is more form than matter. It has, in the history of thought, become a stabilized image for mediating depth and surface, interior and exterior—a space divided rather than vibrated. Except, perhaps, in the case of Bataille, who seems to descend into an enormous mouth, populated by bestial breaths and pulses between humanity and inhumanity.

Bladders, lamellae, and membranes evokes not structure but *aisthesis;* and matter here asserts itself with force. Despite or perhaps because of their minimality, these tissues are sensorially opulent: viscous, tenacious, adhesive. Furthermore, bladders and lamellae are unstable materials, traversed by spasmodic movement, a fibrillation, an opening and a closing. They seem to embody the pulsation of the unconscious, which, as Lacan repeatedly emphasized, operates like a camera shutter, opening and closing intermittently. Bladders and lamellae are materials whose viscosity traps the aura of the universal, of the intensive. They do not let it exist as an external, independent space. They draw it in, retain it, hold it. One senses this virtual drawn into matter in the very attack of the spasm—a point of contact that is also, paradoxically, a point of release.

There are materialities and materialities, after all. Those studied by René Thom, for instance, are topologically stable and repeatable forms—*fold; cusp; swallowtail; butterfly; elliptical umbilic or pyramid; hyperbolic umbilic or wallet; parabolic umbilic or mushroom*—all descending from instability, but nonetheless remaining forms, not materialities. Indeed, they remain indifferent to the substrate of the physical phenomenon analyzed, whether chemical, biological, linguistic, historical, or psychological. It is no surprise that René Thom repeatedly declares himself in service of the continuous, even when probing the zones of instability, crisis, and discontinuity. But near the viscosity of lamellae, nothing can maintain a separate, delocalized existence. Everything is retained, except at the very point of *attack*, which is, paradoxically, also a point of *detachment*.

References

1. Bataille, G.: La Peinture Prehistorique. Lascaux ou la naissance de l'art. Skira, Geneve (1955) [Trans. Prehistoric painting: Lascaux or the Birth of Art, translated by Austryn Wainhouse, Skira, Geneva; Macmillan: London 1980]
2. D'Anna, G.: Introduzione. In: Hartmann, N. (eds.) Ontologia e realtà. Morcelliana, Brescia (2021)
3. Derrida, J.: La dissémination. Seuil, Paris (1972) [Trans. Dissemination, by Barbara Johnson, The Athlone Press, London 1981]
4. Derrida, J.: La vie la mort. Seuil, Paris 2019 (1975–1976) [Trans. Life Death, by Brault, P.-A., Naas, M. University of Chicago Press, 2020]
5. Derrida, J.: Penser à ne pas voir. Écrits sur les arts du visible. Éditions de la différence: Paris 2013 (1979–2004) [Trans. Thinking out of sight. Writing on the Arts of the Visible, edited by Michaud, G., Masò, J., Bassas, J. with new translations by Milesi, L., The University of Chicago Press, Chicago, London, 2021]
6. Di Bella, S.: Aggregatum duorum statuum contradictorium. Mutamento, contraddizione e identità in Leibniz. Fogli di Filosofia **11**(2019-1):23–41 (2019)
7. Gamba, A.: "Leibniz als Metaphisiker": Nicolai Hartmann a confronto con la filosofia leibniziana. Riv Di Filos Neo-Scolastica **106**(1), 131–162 (2014)
8. Hartmann, N.: Der Aufbau der realen Welt. Grundriss der allgemeinen Kategorienlehre. W. de Gruyter, Berlin 1964 (1940)
9. Hartmann, N.: Leibniz als Metaphysiker. In: Hartmann, N. (eds.) Kleinere Shriften. Band 2, Walter de Gruyter, Berlin 1957 (1946)
10. Jorgensen, Larry, M.: By leaps and bounds. Leibniz on Transcreation, motion, and generation of minds. Leibniz Rev **23**, 73–98 (2013)
11. Lacan, J.: Le Seminaire. Livre XI. Les quatre concepts fondamentaux de la psychanalyse. Seuil, Paris 1973 (1964) [Trans. The Seminar. Book XI, The four fundamental concepts of psychoanalysis by Sherida, A. Norton & Company, New York 1998]
12. Latour, B.: Pandora's hope: essays on the reality of science studies. Harvard University Press (1999)
13. Latour, B.: An inquiry into modes of existence: an anthropology of the moderns. Harvard University Press (2013)
14. Leibniz, G.W.: Pacidius Philalethi. In Sämtliche Schriften und Briefe: philosophiche Schriften. Berlin Academy Edition (1676) [Trans. The Labyrinth of the Continuum: Writings on the Continuum Problem (1672–1686). Trans. and eds. by Richard Arthur, New Haven, Yale University Press 2001]
15. Levey, S.: The Interval of Motion in Leibniz's Pacidius Philalethi. Noûs **37**(3), 371–416 (2003)
16. Lombardi, A.: La caverna come immagine del soggetto psicoanalitico in Jacques Lacan. In: L'inconscio. Rivista Italiana di Filosofia e Psicoanalisi, no. 9 (2020)

17. Lorenz, K.: Die Rückseite des Spiegels. Versuch einer Naturgeschichte menschlichen Erkennens. Piper& Co. (1973) [Trans. by Ronald Taylor, Behind the Mirror. A search for a natural history of human knowledge. A Harvest/HBJ book, New York and London 1978]
18. Ronchi, R.: Cioè. L'ontologia di Bruno Latour. Aut Aut **402-2024**, 127–140 (2024)
19. Thom, R.: Stabilité structurelle et morphogenèse. Essai d'une théorie générale des modèles (1972) [Trans. by Fowler, D.H. Structural Stability and Morphogenesis. An Outline of a General Theory of Models. Addison-Wesley 1975]
20. Thom, R.: Modèles mathématiques de la morphogenèse. Recueil de textes sur la théorie des catastrophes et ses applications. Union générale d'éditions, Paris (1974) [Trans. Mathematical Models of Morphogenesis, Ellis Horwood Ltd. 1984]
21. Xenakis I.: Musiques formelles: nouveaux principes formels de composition musicale. In: La Revue Musicale. Editions Richard-Masse (1963)

Chapter 3
FUNDUS ANIMAE

Domenico Licciardi

Abstract This contribution explores the notion of aesthetic experience through the concept of *Fundus Animae*. This concept is examined at a pivotal moment in the history of thought: the transition from Leibnizian metaphysical psychology to the emergence of Baumgarten's and Herder's aesthetics. The aim is to unveil the hidden thread that unexpectedly links differential calculus to the modal ontology of the soul, subtle psychological variations to the *lumen* of the poetic and aesthetic object, and the rationalistic drive toward logical objectivity to the arising of an intimacy irreducibly distant from the subject experiencing it. From this unfolds a reflection on a most delicate form of plasticity: the strange morphology of an infinite *pro-tension* of the sentient being, wholly animated by impure forms of desire.

3.1 Introduction

William Pope.L's *Hospital* stands at the intersection of those practices that Kant grouped under the category of plastic arts: sculpture and architecture. The latter is reverted to its most fundamental activity—neither drawing nor projecting, but carpentry. The piece encompasses a chaotic structure, poised between incompleteness and collapse. Wooden beams intersect with sheets of newspaper, fragments of shattered sinks, and broken toilets. A fine dust of white and gray concrete lies dispersed throughout its foundation. Two wall-mounted shelves support transparent bowls containing other powders. Above them, a caption states: "Dust, sprinkle at will". While acknowledging the call for interactive engagement, we will venture to suggest that the observer always casts a sort of particulate material onto the artwork. The falling dust takes shape, while, shaping something new, a kind of bodily adherence that covers the material work of art. As Duchamp points out: "any form is the perspective of another form according to a certain *vanishing point* and a certain

D. Licciardi (✉)
Università della Calabria, Arcavacata, Italy
e-mail: domenico.licciardi@unical.it

S. Vizzardelli and R. Panattoni (eds.), *Plasticity and Discontinuity*, Lecture Notes in Morphogenesis, https://doi.org/10.1007/978-3-032-03119-8_3

distance" [47, p. 45].[1] There is an inherent geometry of the aesthetic experience, an interplay of perspectival sets (or sets of functions describing perspective lines), whereby form takes shape in the spatial and temporal interval of pre-subjective coordinates. There, in turn, function as images or functional values of a codomain, although the language of analysis has a purely descriptive role in this context. This sensitive, yet phantasmatic set emanates a kind of orientation that anticipate the subject of the aesthetic experience. If we seek to evade succumbing to the metaphysical conception of an eternal present, wherein all that manifests has always been there, even if already veiled our all-too-human eyes, then an inquiry must be posed: how might sensation presage that which is heterogeneous, or originally discontinuous, to itself?

Let us commence by arranging the historiographical elements of our discourse. In 1990, Hans Adler's groundbreaking work on early German aesthetics reconnected Herder's thinking to the epistemological (or "gnoseological") concerns that arose after the exhaustion of Scholastic metaphysics (see [1]). In the first half of the seventeenth century, Wolff and Baumgarten marked a final shift in the history of philosophy. They achieved a systematization of knowledge that left little room for further development, except for subtle refinements in logic, ontology, cosmology, rational theology, rational and empirical psychology. Aesthetics emerged in the wake of the latter, as Baumgarten's *Metaphysics* still attests today.[2] However, as Adler demonstrated throughout his work, the discipline became the catalyst for a series of discussions that transcended the Scholastic system, such as the unfathomable of the soul, the indivisibility of being, and what should be properly defined as "empirical" in psychology. For this reason, "it is challenging to express adequately how all these points had an explosively far-reaching power in relation to [Wolff's] philosophical system" [1, p. 25].

To put it clearly, aesthetics—as the philosophical discipline dealing with the exercise of an obscure, "lower cognitive faculty" [4, p. 77]—is a sensible science of singularities that is incompatible both with Scholastic metaphysics and pre-Kantian systematic philosophy (see [2, pp. 199–202]).[3] As we will see in detail, what is often missing in the common conception of such discipline as a "philosophy of art" is the essential role that Leibniz's epistemology and theory of perception played in its foundation. The German philosopher developed an intensive understanding of sensibility, whose rigor persisted problematically in the *Critique of Pure Reason.* While the *Transcendental Doctrine of Elements* asserts that only the pure forms of space and time characterize transcendental intuition, in the *Anticipations of Perception* it is said that "the understanding can anticipate a synthetic proposition of the sort which

[1] For a reading that overcomes the geometrizing and formalist conception of perspective in Duchamp, see [34, pp. 95-146].

[2] Baumgarten introduced the definition of aesthetics in the First Chapter (Section II) of the Third Part of *Metaphysica* (§533), devoted to *Empirical Psychology* [5, p. 205].

[3] See in particular the conclusive section of the essay, symptomatically entitled: "*De singularibus non est scientia*: L'irréductibilité de l'œuvre d'art". By recalling Aquinas' famous proposition, Adler intends to show Herder's (and more generally early aesthetics') capacity to challenge Scholastic metaphysical assumptions. On the aesthetic notion of simgularity, see also: [51, p. 55].

that concerning the degree of everything real in appearance" [31, p. 294], such as the one stating that there are no empty spaces and times in experience (see also [32, p. 58]). How can the faculty of understanding know a priori that phenomena always contain a certain degree of reality? How should we interpret the notion of "degree"? As Alessandra Campo suggests, the latter functions here as a transcendental instance that anticipates not the (subjectively) objective reality, but the *Sachheit* itself (see [11, pp. 34–35]).[4] "Intensity" expresses *synthetically* the connection between the transcendental subject and the real in sensation. It is no coincidence that the *Critique of Judgment* ponders the impossibility of deciding whether the aesthetic effect of different intensities of visual or auditory vibrations should be attributed to sensation or reflection (Kant ultimately leans *normatively* towards the latter): here lies the indecisiveness between perceived intensity and intelligible difference (or mathematical differential) that characterizes the very concept of "degree in sensation" (see [33, pp. 201-202]).

Firstly, our paper aims to investigate historically an intensive understanding of aesthetics (on this topic, see [29], secondly, it intends to propose a contemporary reading of the aesthetic experience (and knowledge) by building upon such historical background. In order to prefigure our major outcome, we assert that the metaphysical conception of intensive aesthetics is grounded in an ontological continuum entirely justified by divine potency, insofar as it reconciles what is inherently absolute in itself (the monad). Conversely, an intensive aesthetics entirely founded upon the notions of affection, inclination, and desire introduces an element of discontinuity in the genesis of the aesthetic experience.

3.2 "L'essentielle de la Monade… un sombre fond". Sensibility and Virtuality

Let us begin by recalling Gilles Deleuze's interpretation of perception in Leibniz's philosophy, as presented in the seventh chapter of his 1998 essay *Le Pli* (*The Fold*), focused on the relation between the soul and the body.

As Deleuze points out, it is essential for the understanding of the monad to observe that it has a "dark background": "everything is drawn out of it, and nothing goes out or comes in from the outside". Such obscure ground is the most intimate exteriority of the monad, akin to "an obscure object" that "lives in me" [16, p. 85]. This dark background is also the reason for the existence of bodies. What is enveloped demands extension in order to be represented within the universe. This is coherent with Descartes' ontological argument concerning matter as *res extensa*. However, Leibniz's argument is far more articulated. Monads as such have "no parts, neither extension, nor shape, nor division" [39, p. 14]. Yet they represent a universe to themselves. Although the dark ground, as the essence of the monad, will remain inaccessible to representation, the body allows the individual soul to exist in the perception

[4] This topic was carefully analyzed in: [52].

of others. Existence is a play of mirrors. Technically, a reflective layer cannot be seen: we see reflections, but never the surface itself. The body is a reflection, the dark ground is a mirror, but both are actual only for a third perceptive instance: the *other* monad. Hence, Leibniz's "moral deduction of the body" corresponds to the discovery of alterity in the ontological justification of the body as such [16, p. 85].

It should be clear at this point that the issue at hand pertains not only to the relationship between the soul and the body, but also to the ontological nature of the world. At this level, the world exists as it is "included in each [soul] in the form of [...] *present and infinitely minute elements*" or as "an infinity of tiny folds (inflections) endlessly furling and unfurling in every direction" (ibid., p. 87). These folds (*plicae*), as Deleuze says, resemble Simon Hantaï's intertwined stains and "syncopated" rips. Their dynamics of invaginations and evaginations recall Jean Harp's morphogenetic style (see in particular *Constellation According to the Laws of Chance*). Hence, the universe is virtually a perpetually fluctuating motion traversing the tunnels of spiritual wrinkles—something akin to Gaspar Noé's portrayal of the Kabukicho district in *Enter the Void*.

How is the world to be known? As the body materially opposes resistance (ἀντιτυπία) to the darkness of the soul, sense perception limits it formally.[5] Since the vision of the monad consists in its confused (non-distinct) representation of the universe, sensory organs are necessary to limit perception to a sustainable framework, enabling the soul to achieve clarity. Knowledge entails a progressive analysis from obscurity to intuitive perfection, that is, a simultaneous recollection of all the *notae* (marks) composing an object (see [36, p. 23]). Even if the last stage is precluded to human beings and reserved only for God, still the soul needs to perceive clearly and distinctively. This allows for an inquiry into the transition from the virtual representation of the universe to actual perception.

As Deleuze points out, "minute, obscure, confused perception [...] make up our macroperception" [16, p. 99]. The former is made of "*petites perceptions*" [38, p. 38]: tiny perceptions. They constitute a site in the soul that Baumgarten, in turn, names *fundus animae, campus obscuritas, Feld der Dunkleheit* or "field of obscurity" [5, p. 200]. This implies that sense perception is not receptivity but unfurling—a gradual focusing of qualities and forms generated through withdrawing from a larger and confusing virtual background [39, p. 18]. There are, of course, at least two objections to be discussed here. Firstly, such conception may evoke an Idealist framework, as it seems to reduce external reality to the representational force of the soul. Contrary to this interpretation, it implies a rigorous conception of reality, where the logical and ontological premise is that the actuality of the process always precedes the predicaments of a being. One is not seen because she is visible (i.e., possesses the quality of "being visible" as her predicate),one is visible because she is seen (see Plato, 399 BC,

[5] This might be deduced from the definition of the essence of the soul given by one of Leibniz's most acute readers in the past, namely Wolff. See [53, p. 45]: "*Essentia animæ consistit in vi repræsentativa ūniversi situ corporis organici in universo materialiter & constitutione organorum sensoriorum formaliter limitata*".

10c).[6] Here, it is not the subject, but rather the *actuality* of something that precedes the attribution of predicates. Secondly, tiny perceptions are never motionless. They constitute a dynamical domain, akin to a highly confused state function, resembling a sort of Brownian kineticism. However, while sense perception is not constituted by random phenomena, randomness still characterizes the dynamics of *petites perceptions*. While this may seem contradictory, one should not lose sight of the difference between the virtual and the actual. Constructing the first based on the second amounts to eliminate such difference. To conceive of virtuality as a structured replica of the empirical may seem reasonable, but it implies that experience is ensnared in the web of fixed categories and reason is confined by the constraints of common understanding. However, starting from 1900, Physics has accustomed us to the idea of a scalar difference between quantum reality and the empirical plane of classical mechanics. For instance, planetary orbits are deemed regular, to the extent that they can be described through Kepler's equation. However, this does not mean that all orbital motions are homogeneous. At the subatomic level, the electron is described by a wave function representing a probability distribution, thereby rendering its position indeterminable. The notion of a distinction between a probabilistic microscopic plane and a mechanistic macroscopic plane is not new, and indeed aligns with the logic expounded by Leibniz. Having said that, to further develop an interpretation of *petites perceptions*, we will endeavor to supply a historical and logical background.

Leibniz introduces the concept of *petites perceptions* in the *Nouveaux Essais sur l'entendement humain*, a long answer to John Locke's *An Essay Concerning Human Understanding*. In his view, contrary to Locke's assertion, the mind is not a blank slate but rather akin to a veined marble block predisposed to becoming a statue. He states that the "veining" of the mind corresponds to a manifold of *virtualités naturelles* or "natural potentialities" [38, p. 6]. These latter anticipate conscious perception (the author recalls the Stoic term πρόληψις) and they work as the insensible ground of conscious sensations. This explains the statement according to which *petites perceptions* in psychology (or "pneumatology") are comparable to the "insensible corpuscle" in Physics. Tiny perceptions allow us to investigate the relation between the mind and physical motion. In fact, their theorization avoids all the heuristic difficulties involved in defending empirical observation while considering sense perception as a source of accidental errors. From a logical standpoint, tiny perceptions are postulated in light of "the Law of Continuity", according to which "a leap from one state to an infinitely different one cannot be natural" or, in other words, "*nature never makes leaps*" (ibid., pp. 9–11). If the principle is correct, it is not possible for perception to totally and accidentally diverge from the physical world. Hence, the question underlying Leibniz's argument is: how is sense perception coherent with the perpetual "*variations actuelles*" (or actual variations) of bodies, which can be treated as mathematical dispositions ("*Estres Mathematiques*")? [37, p. 563]. His answer lies in the

[6] "Is what I want to say clear, Euthyphro? I want to say this, namely, that if anything is being changed or is being affected in any way, it is not being changed because it is something changed, but rather it is something changed because it is being changed; nor is it being affected because it is something affected, but it is something affected because it is being affected" (trans. 2002, p. 12).

idea of unconscious perception that relates to conscious perception just as particles relate to physical phenomena, that is, according to a genetic and differential relation. The principle of continuity implies that sensory qualities, forms, and objects, do not appear before our sense organs as accidental externalities, nor are they projections of our mind. Rather, they are phenomena *directly* extracted by the constant flux of physical becoming, which, in turn, can be traced back to the virtual representation of the monad. One may notice here a first, essential connection between Leibniz's philosophy and aesthetics as a *science* of sensible cognition: it is the necessity of the sensible that is at stake. Early aesthetics is inconceivable without this theoretical background. The question does not concern the dogmatic generalization of sensation; rather, it regards the underlying and genetic processes that connect sensibility and the world (such a connection being a part of the general *nexus rerum*). In Leibniz, both *petites perceptions* and physical particles consist of "insensible variations" and pertain to a logic in which the difference among individualities (according to the principle of indiscernibles) is "more than numerical", that is, infinitesimal [38, pp. 9-10]. Both the physical world of experience and sensible phenomena spring out from intensive processes that can be described in terms of differentials.

Hence, the question "of knowing how we move from minute perceptions to conscious perceptions" demands that we understand "literally—that is, mathematically—that a conscious perception is produced when at least two heterogeneous parts enter into a differential relation that determines a singularity. It works thus in the equation of circumferences in general: $\mathrm{ydy} + \mathrm{xdx} = 0$, or $\frac{\mathrm{dy}}{\mathrm{dx}} = -\frac{\mathrm{x}}{\mathrm{y}}$ expresses a determinable magnitude" [16, p. 88]. The foundation of the process that leads from microperceptions to macroperception based on the differences among heterogenetic instances must not confuse the planes of discourse: it is true that the principle of continuity justifies the existence of obscure perceptions, but nothing in this principle makes the realm of small perceptions homogeneous. Continuity clearly concerns the relationship between the physical world and the perceived world, not the relation among small perceptions. The intensive conception of sense perception suggests that the latter arises from the relation between variations in the unpredictable dynamics of *petites perceptions*. Such heterogeneous elements generate intensities that traverse unconscious thresholds, finally reaching the field of consciousness. In a broader sense, the possibility of describing sense perception through differential equations suggests that sensibility can be conceptualized through a dynamically nested folding model. In Newton, the fundamental problem of differential calculus is: given the relation between two variable quantities (*fluentes*), what is the relation between their related rates of change (*fluxiones*)? (See [12, pp. 93–95]). However, in this context, due to the relational value of each algebraic function, every variable operates both as a *fluentes* and as the rate of change of other variable quantities. In this perspective, conscious sensation is not a given (*data perceptiva*) but rather the outcome of unconscious, intensive phenomena whose variations are interchangeable and thus virtually unpredictable. Overall, sensibility can be described as an infinite production of becoming whose foundation is composed by singular variables.

In this manner, we may gain insight into an intensive conception of sensibility that is coherent with physical phenomena. However, as we have seen, tiny perceptions can only be postulated according to the principle of continuity. To clarify his argument, Leibniz often refers to the difference between the unitary sound of the ocean and its underlying multitude of waves. Anyway, even a good number of examples does not define an autonomous field of study. In order to obtain a tangible understanding of the relation between virtuality and actuality in sensation, one must emphasize the "remarkable" character of those conscious perceptions that stem from the unconscious field of *petites perceptions*. In reaching this point, we already are within the domain of aesthetics.

3.3 "Vividum dicimus, in quo plura varia...". Lively Perceptions and the Field of Obscurity

In Leibniz's philosophy, the transition from tiny perceptions to macroperception lacks both integrality, in which parts are added to constitute a whole, and organicity, where the whole functionally overcomes the mere integration of its parts. As evidence, one might overlook either microscopic or macroscopic perceptions, for instance, due to distraction (see [16, p. 87], [38, p. 7]). The flaw in conceiving perception according to a relational logic of parts and the whole lies in the fact that both are structurally predetermined. Ultimately, every generative process emerges as posterior to pre-existing instances. This initiates an infinite regression in the order of filiation of parts and the whole. On the contrary, generation entails modal ontology, in order to describe the process of actualization of the virtual. In the context of our inquiry, such logic specifically concerns the transition "from the *ordinary* to what is *notable* or *remarkable*" [16, pp. 87–88]. This constitutes one of the most important themes in Baumgarten's aesthetics.

It is widely acknowledged that the latter is conceived as analogous to Logics. Both disciplines contribute to the inquiry into the inner connection of things (*nexus rerum*), which ideally constitutes the utmost achievement of metaphysics in the Leibnizian-Wolffian tradition. However, while Logics, or *philosophia rationalis* aims at improving the higher parts of the cognitive faculty, Aesthetics, in turn, aims at the perfection of sensible knowledge (see [6, p. 6]). In order to attain this objective, Baumgarten needs to identify the distinctive epistemological mode of aesthetic inquiry.

In Leibniz's epistemology, clarity and distinctness are not necessarily linked, as they happen to be in Descartes. One might have both clear and confused ideas of an object. This occurs when its sensible properties are perceived but one is unable to enumerate them. For instance, judgments concerning art and poetry rely on a *je ne sais quoi*: the object is felt but its beauty is not attributed to any particular quality (see [36, p. 24]). By drawing upon clear and confused perception, Baumgarten theorizes two modes of clarity: "Greater clarity due to the clarity of notes [*notae*] can be called *intensively greater clarity*, while greater clarity due to the multitude of notes can be called *extensively greater clarity*" [5, p. 204]. These modes are not meant to oppose or contradict each other. Rather, proceeding from the comprehensive framework of Leibniz's epistemology, Baumgarten identifies the epistemological degree that specifically corresponds to the sensory liveliness of perception. Therefore, aesthetics concerns art and poetry not because they constitute an arbitrarily established field of inquiry, but to the extent that they are capable of producing vivid, or *remarkable*, perceptions.

"Remarkable", "lively", "vivid", indicate that the individual object is "*omnimode determinatum*", that is, represented at the highest degree of determinacy [4, p. 55], [51, p. 43]. More specifically, a remarkable representation is one that confusedly contains the largest number of marks (*notae*). However, there is an important aspect to notice here. Leibniz points out that clear and confused knowledge alone is insufficient "for differentiating a thing from others" [36, p. 24]. For Baumgarten, this incapacity to identify the object does not affect the possibility of philosophically treating its determinacy. What is at stake is, therefore, a concept of individuality that is indifferent to identity or identification. Hence, from a logical perspective, the distinctive object of aesthetic inquiry is not a singularized identity but rather an individualized singularity. While this difference may seem like a mere metaphysical subtlety, the difference between identification and individuation carries significant implications, as it introduces the idea of a perceived singularity that, insofar as it is independent of the binary logic of identity and difference, cannot be subsumed under more general concepts.

As it is evident, the understanding of subjectivity in aesthetics cannot be the one that is at stake in epistemology. In this scenario, Baumgarten's notion of *impetus aestheticus* describes a subjective condition in which an "excitable nature" is gripped by a state of mind wholly dedicated to the activation of its previously dormant cognitive faculties, enabling the subject to create something as if her mental powers were "raised to the power of two" [6, p. 35]. It is interesting to note that, in interpreting the concept of aesthetic impetus, Niklaus Largier employs the term "plasticity of the soul". In support of his reading, Largier traces the concept of *fundus animae* back to "medieval and early modern theology and mysticism", in which the soul in general was intended "as a realm of possibility" to be actualized or individualized [35, pp. 536–537].[7] However, as a part of it, the *fundus animae* (the background of the soul) remains unactualized. This difference between the virtuality of the soul and its actualization helps to think the notion of plasticity at stake here. On the one hand, by positioning the *impetus* at the core of aesthetics, Baumgarten reinstates the notion of "experience" as "the manner in which the soul assumes shape under

[7] In order to avoid an overly emphatic connection between aesthetics and mysticism, one should notice that in Baumgarten the concept of *impetus* is completely secularized. The examples that he brings about to explain the *impetus* are not only linked to art, but also idleness (*Aesthetica*, §84), drunkenness (§§85–86), and sexual desire (§87). One could notice that even sports can cause such an occasion of impetus; for instance, when Nan Shepher writes, about climbing the Scottish Cairngorms: "It is, of course, merely stupid to suppose that the record-breakers do not love the hills. Those who do not love them don't go up, and those who do can never have enough of it. It is an appetite that grows in feeding. Like drink and passion, it intensifies life to the point of glory". Shepherd [48, p. 6].

sensual influences" (ibid., p. 538). On the other hand, the virtual soul appears as the creative instance that renders ontogenesis a process that is always open to new and unexpected variations. To comprehend this one should return to the very roots of subjectivity.

In *Metaphysics*, we read that "if there is something in a being that can be conscious of something, that is a *soul*" [5, p. 198]. Since thinking itself is an alteration of the soul, "I think" means that "I am modified". In this sense, the soul is a power and thoughts inhere in it as accidents inhere in a substance: "If accidents inhere in a substance, then there is some ground of inherence [...], or *power*" (ibid., p. 136). This means that the soul as virtuality (*vis* or *virtus*) is individualized by its thoughts as the substance is determined by its accidents or the subject is particularized by its predicates. The process of individuation of the soul proceeds from the general to the particular and the singular (see ibid., pp. 198–199). However, what Baumgarten terms *fundus animae* is ambiguous: on the one hand, it merely gathers the obscure and indistinct perceptions of the soul, on the other hand, it is the condition of possibility to distinguish something from all other objects of thought: "if one did not at all perceive the notes of something confusedly thought, then through these one would not be able to distinguish the thing confusedly perceived from others" (ibid., p. 199). This ambiguity is resolved as soon as one recalls that, in Baumgarten as in Leibniz, on a metaphysical level, the process of individuation coincides with the unfurling of the power of representation. Hence, the discourse demands (for those who come after Kant) a double perspective. It is necessary to make the effort to consider the simultaneity between: 1. The modal actualization of the substance through its accidents (from the general to the particular); 2. The intensive representational process from obscurity and confusion to clarity and distinction (see ibid., p. 200). In this perspective, the *fundus animae* is simultaneously the field that precedes all other degrees of representation and the instance of its own actualization. To put it differently, the background of the soul is both the first correlative of the initial act of thinking and the first movement in the process of individuation of the substance-soul as such.

This helps to interpret the relationship between aesthetics and subjectivity in Baumgarten. On the one hand, by engaging the ground of the soul (*fundus animae*), the aesthetic *impetus* reinstates the temporality of virtual perceptions, whereby the past, the present, and the future deviate from their hinges (see [6, p. 36]; compare with [38, p. 8]). On the other hand, what is remarkable is thus because it confusedly gathers the greatest number of marks (*notae*) within a single perception (*claritas extensiva*).[8] The aesthetic experience emerges from the heterogeneous relationship between the indeterminacy of the soul and the utmost determinacy of the object of art or poetry. The artwork, as otherness, as what is *exterior* to the process of individuation

[8] Compare: [5, p. 201] ("The more motes a perception embraces, the stronger it is [...] Hence, an obscure perception comprehending more notes than a clear one is stronger than the latter, and a confused comprehending more notes than a distinct one is stronger than the latter. Perceptions containing many perceptions in themselves are called *pregnant perceptions*") and Id., 1735, trans. 1954, p. 43 ("The more determinate things are, the more their representations embrace. In fact, the more that is gathered together in a confused representation, the more extensive clarity the representation has, [...], and the more poetic it is");

of the subject, instantiates a process that unties what is bound and makes room for new, unstable individuations that are independent of the self-identity of the subject. Aesthetics and subjectivity connect at the level of what Deleuze interpreted as the most intimate exteriority of the soul (the dark ground of the monad or the soul).

In this context, Herder will develop a conception of aesthetics that is inseparable from an inquiry into the genesis of subjectivity. However, Herder's peculiarity manifests at a heuristic level: he does not attempt to reconstruct the process of subjectivation as a chronological sequence, but rather to destroy the philosophical structures that conceal the sensible genesis of all processual becoming from virtuality to actuality.[9] This leads us to consider his work as an aesthetic and critical genealogy of subjective plasticity. However, this destruction can happen only *through* the work of art, which bears a sublime alienation emerging, at this time, at the very heart of similitude: the statue as an anthropological element that, coming to life, reveals the utmost exteriority of the human being from itself. Following this, aesthetics and the philosophy of subjectivity appear as inseparable.

3.4 "Da ist seine Bildung: die Bildung eines Körpers". Aesthetics and Subjective Plasticity

While Herder's thought is notoriously challenging to categorize, it is undeniable that Leibniz and Baumgarten had a significant influence on the development of his psychological reflections and aesthetic philosophy. However, understanding the role of art and sensibility in this context necessitates advancing some epistemological considerations. These three branches coalesce as follows: faced with the paradox of being (epistemology), Herder's exploration of the genesis of the soul (metaphysical psychology) leads him to envision the arts as the field of inquiry for a new perspective on the constitution of subjectivity (aesthetics). Let us delve deeper into this connection.

The epistemological paradox of ontology consists in the fact that existence is the necessary foundation for every possible metaphysics, yet it remains inherently obscure, that is, inaccessible to knowledge. Based on the gnoseological equivalence between obscurity and sensibility in Leibnizian and Wolffian epistemology, Herder concludes that being is "the most sensory of concepts" [23, p. 60]. "Sensible" and "obscure", understood as epistemological properties of existence, signify that the latter is "*unzergliederlich*" [23, p. 9]. This word ("indivisible") refers to the unity of

---

[9] See on this point a note by Herder entitled "*Negative Philosophie*", quoted in Heinz, Clairmont, 2009, p. 44: «Erst muß m[an] z[er]stör[en], denn [au]fbauen, m[an] z[er]störe alle Systeme d[urch] ei[ne] negat[ive] W[issenschaft] u[und] führe alsdenn [au]s dem subjekt[iven] Princip[ium] e[ine]s [au]f, w[a]z ganz wenig behauptet, d[die] Grade d[er] Gewißh[eit] bei jedem Satz bestimt: d[ie] Art d[er] Demonstrat[ion], und ihre mögl[iche] und wirkl[iche] V[er]schied[en]h[eit] [au]s e[in]and[er]setze:—unt[er]suche, w[ie] d[ie] Wißenschaft[en] alle V[er]sch[ie]d[ene] s[in]d nach Ihrer Entsteh[un]gsart:—subjekt[ive] Art des Denkens;—objekt[ive]. Besch[affenheit] u[nd] Methode».

being as unanalyzable (Herder is usually very attentive to the etymology of words, and "ἀνάλυσις" means "division"). A thing can be analyzed down to its minutest details, starting from a mere listing of properties to the logical analysis of its possibility according to the non-contradictory inner connection of its predicates. However, the analysis inevitably reaches a *non plus ultra*, an unanalyzable "*Klumpen*" (ibid., p. 11), namely the fact that the thing *exists*. Our knowledge can only be certain about the external relations of an object, but the internal connection of the thing, which determines the *reason* of its existence, is inevitably speculative. For example, the multiplication of epithelial cells on a substrate indicates that each of them tends to occupy the vacant spaces on the surface. However, based on this alone, it is not possible to determine whether the distribution is a purely mechanical process or serves a specific purpose.

Nevertheless, the attribution of existence must inherently contain some degree of certainty, unless one wishes to lapse into skeptical nominalism. Consequently, the source of such certainty must be sought in a middle term between the mind and the world. According to Herder, we are certain of the existence of being because we are somehow supplied with "the impression of its nature" [23, p. 61]. In this aftermath, it is possible to discern real existence (*Realsein*) from logical being (the notion that metaphysical ontology logically deduces from the concept of *Nihil* or nothingness), in order to refuse a conception of the latter as mere *complementum possibilitatis*—that is, to the force of actualization of all that is virtually possible (see [5, pp. 100–103]).[10] In the same vein, it is further possible to distinguish real and logical possibility (*Real-* and *logische Möglichkeit*). By focusing on such modal ground, what Herder seeks is the "real" (i.e., not merely logical) unfolding of possibilities from the actuality of existence. Taking a closer look, this justifies Herder's interest in individual psychology and general history. His relentless emphasis on genetic phenomena (a mark of his typically genealogical style) is rooted in the conception that there is no way to comprehend the nature of something other than grasping a *deviation* within the fabric of existence, which in turn corresponds to the moment at which the former came into being.

From this viewpoint, there is no other way to grasp the essence of subjectivity other than moving back to the origin of the soul, where the "impression of being" can be considered as an inner sense: the mere awareness of one's own existence. We can specify this point through exposing one of the most important differences between Leibniz and Herder. By relying on the concept of "*Innern Sinn*", Herder challenges the Leibnizian model of the mind as a veined block of marble, asserting instead that nothing innate or pre-existing to the mundane subject exists but a "bud" [28, p. 236]: a *Leimengebäude* (a glue-trap structure), a (plastic?) composite of soul and body generated from a series of complications within the physical dynamics of the material universe, at the end of which a perceptible feedback (*Rückwirkung merkbar*, translated into English as "*reaction*") causes the soul to bring the manifold of sensations into the unity of apperception (ibid., p. 208). However, the relevant

[10] Herder develops this argument as a pupil of Kant. On Kant's critique of Baumgarten's argument, see: [30, pp. 107–110].

point in this regard is that Herder does not proceed from the purely immaterial force of representation, but from a soul-body composite (one may recall the Aristotelian σύνθετον, ἐκ ψυχῆς καὶ σώματος συγκείμενον) that, however, is not formed but formant, not derived but genetic. Following this, he envisions a new conception of psychology. Inspired by Haller's theory of irritability, he advances the notion that the metaphysical investigation of the monad should be supplanted by a physiology, or even a physics, of the soul (see [42, pp. 53–97], [50]. This shift also corresponds to the temporalization of the *fundus animae*, which is no longer intended as a timeless "field" but rather, more dynamically recognized as the genetic moment of sensitive and empirical faculties (see [40, p. 95]).[11]

Hence, the inner sense is not a fixed ground but the correlative of a soul-body composite that actualizes through space, time, and force in sight, hearing and touch.[12] In this manner, he draws from Wolff (but revisited through Crusius and Kant) a genetic sequence that allows conceptualizing the individuation of the soul (and the body). However, this sequence is unable to explain why our "lower faculties" are more potent than the "highest" ones. In order to justify this, one should take into account that the former "develop earlier" [25, p. 193]. What is genetic, what is closer to its springing into existence, possesses greater potency, determinacy, and vividness, while what follows in time exhibits a decrease in intensity and a tendency towards formal abstraction.

What is powerful in origin is the field of *petites perceptions*. Here, Leibniz is effectively interpreted through Rousseau[13]: "the whole ground of our soul consists of obscure ideas, the most vivid and most numerous ideas, the throng from which the soul prepares its more refined ones; these obscure ideas are the most powerful mainsprings of our life, make the greatest contribution to our happiness and unhappiness" [25, p. 193]. Here we find also a more defined logic of the entanglement between a modal and a relational logics: the process of actualization tends to unity (the unity of

[11] We take the opportunity for a consideration of the concept of plasticity, which subtly comes into play here as soon as we recall its use by Largier (on which we will return shortly in connection with Herder). As Catherine Malabou emphasizes in her commentary on paragraph 258 of his *Enzyklopädie der philosophischen Wissenschaften im Grundrisse*, plasticity is a term capable of describing not only the dialectical relationship between the substance-subject and its accidents (a question on which Malabou constantly returns, up to the *Ontology of the Accident*), but also the reversal that metaphysics performs from the punctual and Aristotelian conception to the synthetic-transcendental (Kantian) conception of time. The latter point is certainly the most decisive, as it reveals the deconstructive instance of the vulgar conception of time within Hegelian dialectics. However, if we observe the aforementioned temporalization of the *fundus animae* as the first moment of the actualization of the substance-soul (Baumgarten), we will obtain a nucleus, namely a temporalization "in embryo" of substance itself, not entirely disconnected from Herder's subsequent anti-metaphysic turn towards a philosophy of history. This effectively opens up to the possibility of a new historical-philosophical investigation into plasticity (regardless of its semantic use), since if aesthetics is the birthplace of the concept, it must contain, in the strict terms of Hegelian dialectics, the instance of its logical-ontological development [41, pp. 13–18].

[12] Herder conceived of this threefold and original connection since the 1766–68 manuscript entitled *Plato Sagte: daß unser Lernen bloß Erinnerung sei*, which Marion Heinz excellently analyzed in: [21, pp. 43–80].

[13] On Rousseau's influence on Herder's conception of individuality and subjectivity, see [21, p. 68].

the soul) but its genetic elements (obscure ideas) are a multiplicity of heterogeneous elements. Such *dunklen Ideen* are connected to sensation in the following way. The "self" (a merely conceptual unity that gathers the manifold of obscure ideas, but that more nominal approaches could easily reduce its genetic plurality) develops during the early stages of life and sensation plays a crucial role in the process: it molds and sculpts the soul potently. Here, "sensation" is not the empirical correlative of a receptive faculty but the chisel through which nature shapes the subject. From the inception of existence, the undetermined soul-body composite is subjected to the plastic force of external stimuli: "with repeated, identical sensations the infant forms his first judgment", which in turn is "obscure": "an inner feeling [...] preserved as sensation" (ibid., p. 194). The judgment "preserved as sensation" is an unconscious aggregation of heterogeneous elements[14] that secretly nurtures self-identity, as it constitutes the true source of individuality: "If no two human souls are wholly alike, then perhaps in their natures, too, there may be infinitely varied and modified combination of powers, all of which can yet possess an equal degree of reality as an aggregate" [25, p. 194]. But the aggregate is not defined once and for all. It never ceases to move or undergo modification. It functions as the metastable instance of subjectivity, similar to a continuo bass capable of radically altering the atmosphere of a melody, or as a chasm perpetually on the verge of *self*-collapse. Aesthetics' collateral effect reveals the "abyss" at the "origin" of the "soul" (ibid., p. 195).

In *Plastik*, Herder reexamined the distinction between tactile and visual experience[15] to differentiate between the unique aspects of sculpture and painting. However, due to the same logic that we observed above there is no way to understand a form of art but to retrace it back to its origin. According to Herder, the autonomy of sculpture emerges when it evolves to depict anthropomorphic figures. While there is undoubtedly an anthropological dimension to this perspective, what strikes him is not the self-recognition of humanity in the artwork. Rather, he advances the idea of an impression of terror, shuddering, and awe, stemming from the primitive feeling that the "the work appear to *move* and *come to life*" [27, p. 91]. The reference to Daedalus' automata expresses a kind of "demonic" life, evoking an ambiguous reaction of repulsion, wonder, and reverence, a feeling connected to the infinite colossal as one of the categories of the sublime (see [10, 36, 3], [46, 2, VII-VIII]).

This primeval feeling resonates in the modern (i.e., aesthetic) experience of art, in which "the sculpture lives" under the amazed touch of the observer [27, p. 41]. As Largier points out, such experience is characterized by the fact that "all sensation turns into touch—and touch encompasses virtually all sensation" [35, p. 538]. In this manner, sculpture "un-builds" the *Bildung* (formation), that is, the intellectual dominance of objects through the exercise of the senses that Herder reconstructs in the first chapter of the essay [27, p. 37]. The emergence of touching (*Gefühl*) as the fundamental epistemological act regains its priority, enveloping the scene once more.

---

[14] Visual, tactile, auditory *impulses*, for at the level of the inner sense the separation of object and subject is yet to be established. See [21, p. 61].

[15] George Berkeley attentively analyzed the difference and relation between the two senses, reaching the conclusion that: "Visible figures are the marks of tangible figures" [8, p. 60].

It is at this moment that "the *transposition of our entire human self* into the figure" of the statue occurs (ibid., p. 78). The plasticity of the soul is not the shaping of the latter through external influences, but the *relocation* of an undetermined ground (indeed the soul) in another aggregate, configuration or form.

But what does it mean that tactility regains its priority? What does *Gefühl* really signify? As Elena Agazzi points out, Herder retraces the notion to its Latin etymology (*cum tangere*, *contingens*, "contingency").[16] Here, the concept does not refer to "accidentality". Rather, it should be read as a Neoplatonic concept expressing the unlimited pervasiveness of being throughout all that is contingent, or the irreducible repetition of unity in multiplicity.[17] *Gefühl* refers to a virtual tactility that is irreducible to the conscious sensory act of physical or empirical touching. When Herder utters his notorious formula, "*Ich fühle mich! Ich bin!*" ("I feel myself! I am!"), he does not seek to oppose the immediacy of feeling to the abstraction of reflection. One should interpret the sentence in light of the idea that the genetic moment of individuation proceeds by the shaping of a desiring "*Nervenbau*": a system of living fibers animated by the natural forces of attraction and repulsion [28, pp. 187–188]. According to Herder, if souls are conceived as physical phenomena, the possibility of self-cognition would be precluded, as the individual would endlessly collapse (*fallen*) inward upon itself [24, p. 229]. In a manner radically different from an entire tradition stemming from Descartes to Baumgarten, for Herder, subjective thinking is not genetic but rather an after-effect. Instead, the latter must be sought in the individuation of the soul, which in turn occurs through sympathy. In this perspective, "sympathy" specifically signifies sensuality as enjoinment (*Wollust*). Originally, the soul extends its body because it longs for the *other*. The body is not the envelope of the soul but its form, its actual existence. Enjoyment is the primeval plastic force (*Kraft*) that molds the genetic *Leimengebäude*, the primal soul-body composite, well before the emergence of a subject capable of saying "I think". In this wake, aesthetic is disruptive because, by "touching" the notes of enjoyment, it operates at the level of the pre-subjective plasticity of the subject. Every experience of art entails a trans-position (*transponere*), a departure from shape to shape,[18] a "contingent" metamorphosis provoked by a movement of "destruction" and reformation of a metastable aggregate: the *fundus animae*.

---

[16] See Agazzi's comment to Herder's *Zum Sinn des Gefühls*, in [28, p. 847, n. 1].

[17] On Herder as the foremost expert on Neoplatonism in Germany prior to Schleiermacher's translation of Plato's *Dialogues*, see [18, p. 162].

[18] This prompts consideration of a potential, latent connection between aesthetics and creative mathematics. In *The Prose of the World*, Merleau-Ponty ponders Carl Friedrich Gauss's formula for the sum of natural numbers. What occurs in the transition from the series 1, 2, 3... n to the formula $\frac{n}{2}\text{x}(n+1)$? For the French philosopher, the latter represents a *Gestalt* articulated through an act that reorganizes the initially provided numerical series, which is not purely analytical but creative. This seems to suggest that mathematical development unfolds through configurations and reconfigurations where the symbol serves as hyletic perception, and the mathematician's act is immanent to the transformations occurring in the realm of formalization and symbolization of mathematical knowledge. In this context, both mathematical and aesthetic creation reveals a formal, or virtual, kinship, consisting in an immanent transformation of shapes or *Gestalten*. See: [20, pp. 105-107], [44].

3.5 "Plus que numérique". Aesthetics and Becoming

Now, one of the most significant achievements of Herder's aesthetics is that, lacking a fixed perspective, the experience he describes is "more than numeric", for when it comes to sculpture, "a thousand points of view are not sufficient" [27, p. 41]. Marco Mazzeo's insight regarding geometry as the synesthetic common ground of sight and touch in *Plastik* allows for a distillation into Herder's perspectivism to its geometric framework (see [43, p. 50]). The infinity of viewpoints delineates a "circle" in which the distance between two diameters (considering the outlined triangle as specular to a symmetrically opposite figure with the same diameters on the line of the radii) forms a non-metric distance, or a relationship of infinitesimal closeness [27, p. 41].[19] The concept of infinity in Herder's aesthetics bears little resemblance to the traditional understanding of the sublime. Despite its reference to the colossal, it is not the latter that determines the endlessness of the experience, but rather the modulation of form itself.[20] As Andrew Benjamin states, in Herder, touching extends endlessly. Its temporality is characterized by the perpetual unfolding of "always more", and its ontological status is one of a process directly emerging from matter—"matter as mattering [...] being as mattering" [7, pp. 246–248]. "Plastic" is not only the quality of the artwork to geometrically extend into space. It also denotes the unstable and material difference among heterogeneous variations (differential variations among, not only the eye and the hand, but also space and form, surface and body, and so forth).[21] Seen materially, sensibility reveals its most intimate problematic nature: there is no sensible singularity without an equally sensible background, constituted of other singularities, and consisting of intensive relationships. Duchamp attempted to isolate the infinitesimal element of sensation through glass plates (see [9, p. 26], [17, 19]). However, in general, the problem remains as the irresolvable tension between background and figure.

In this perspective, art is the domain of such specific problematization of sensibility. This emerges clearly in the second half of the nineteenth century, when Cézanne transformed the background from a mere backdrop into a visual challenge in its own right. Notably in Van Gogh, and even in some of Gauguin's panoramic paintings (such as *Above the Sea* and *Haystacks in Brittany*), each figure emerges dynamically, struggling against forces within the painting (cutting colors, incisive panoramic outlines, bursts of light). These ones make it nearly impossible to maintain a profound distinction between the foreground and the background. Figures cannot stand, nor

---

[19] In trigonometric terms: the infinitesimal ratio (or the ratio between two quantities as they proportionally tend to zero) between the sine and cosine of an angle associated with a tangent. On this argument, see: [12, 14, pp. 103-106].

[20] See [27, p. 94]: "The hand can rest on the sphere and the sphere in the hand. But with articulated forms, the work of the hand is never complete: it goes on feeling, so to speak, *infinitely*. This is true above all of the form of the human body, even when it appears on the smallest crucifix".

[21] It is in the wake of a differential relation that we interpret the constant oscillation between touch and sight in Herder's aesthetic of sculpture. On the relationship between these two senses in *Plastik*, see [13, pp. 95–96].

may blend: they are forever consigned to an unsustainable imbalance. Cinema underscores this instability through the techniques of rotoscoping or animated painting, as in Richard Linklater's *The Waking Life* or even in the Welchmans' *The Peasants*, where the dizzying tremors of the image persist, never settling but in the darkest or brightest scenes (where there is hardly any difference between the background and the figures). The stakes of a differential aesthetics are played out in the interplay of intensive forces connecting the *fundus* and the form. In other words, it concerns precisely the difference between virtuality and singularity in the domain of the sensible.

When Deleuze emphasizes Francis Bacon's endeavor to shift the focus of painting from the relationship between form and matter to the one between materials and forces, he does not appear to express a desire to relinquish the issue of form. Rather, he states the necessity of reframing the question within a different context. The notion of resonance demonstrates the necessity of substituting a hylomorphic logic of plasticity with a metastable conception of individuation—as Gilbert Simondon asserts, "the being's internal resonance is the tension of metastability" [49, p. 226]. One could observe that the application of Newtonian laws to the motion of materials under the influence of external forces (a process that relevantly inspired Simondon and, indirectly, Deleuze) can be represented through differential equations describing the system. While art is undoubtedly distinct from physics, sensations in the artwork resonate, revealing a rhythmic frequency that precedes and informs the morphogenesis of sensible qualities and forms.

As Deleuze points out: "it is a characteristic of sensation to pass through different levels owing to the action of forces. But two sensations, each having their own level or zone, can also confront each other and make their respective levels communicate" [15, p. 65]. In wake of its "intensive reality", sensation essentially consists in vibrations always subjected to "allotropic" variation (ibid., p. 45). This means precisely that sensation can exist on different levels at the same time: "it is each sensation that exists at diverse levels, in different orders, or in different domains. This means that there are not sensations of different orders, but different orders of one and the same sensation". Given this quality of sensation, resonance is more complex, since it results from the meeting of two or more vibrations as their amplification surpasses their peculiar plateau. Consistently with what we have seen until now, this process can be seen as the emergence of sensation, namely the Figure, as it stems from a virtual multiplicity constituting its genetic background: "It is the nature of sensation to envelop a constitutive difference of level, a plurality of constituting domains. Every sensation, and every Figure, is already an "accumulated" or "coagulated" sensation, as in a limestone figure" (ibid., 37).

However, if each sensation can draw towards the other (with the latter being the elemental process in the constitution of the Figure), this happens for two reasons: folding and appetition. Each of these concepts warrants separate considerations.

As for the first, while resonance denotes the relationship among sensations on a canvas at a microscopic level, both Figure and the plasticity of the soul pertain to the interaction between the artwork and the observer.[22] One could argue that the aesthetic relationship (and consequently, experience) exists because veins or folds are ubiquitous. They are everywhere: in matter, in marble, on canvas, as well as on the surface of the brain, in memory, and even (according to the Leibnizian language of Baumgarten and Herder) in the soul. Far from meaning that they constitute a homogeneous continuum, one should remember that intensity arises as they differ from each other. However, what is important here is that the virtual and the sensible differ ontologically, yet from an aesthetic perspective, their distinction is inversely proportional to the abundance of the folds. Extensive clarity leads to amazement and vertigo (*impetus aestheticus*), which is precisely why it resembles the monadic perception of the universe. From an aesthetic point of view, the difference between the virtual and the sensible *may be* very subtle, up until the closest point of asymptoticity.

The second question concerns desire. If resonance in the artwork is not provoked by the application of an external force to a physical body, what is its "cause"? In Leibniz, appetite is an essential attribute of monads, enabling change in perception [39, p. 16]. It roots in their ontological nature, that is, in their finitude. Souls perceive because they desire, and they desire because their reality is imperfect or incomplete (it lacks total, or perfect actualization). Through their *vis repraesentativa*, souls are astonished by the universe, but they wish more. They unite, intertwining intensely to give rise to sensory qualities and forms, yet their desires or inclinations yearn for more. Their appetite is endless, yet God completes the unfathomable circle of inclination, movement, change and generation. Only in him, actuality and possibility are perfectly simultaneous. He is the one who harmonizes reality and appetition, causality and desire, such that "every movement has some effect on distant bodies in proportion to their distance" (ibid., p. 26). But divine wisdom irremediably surpasses the gnoseological capacities of human beings. Therefore, one could notice that, even if knowledge is an endless process, the idea that everything is connected (*nexus rerum*) is all that the soul needs to know to satisfy its metaphysical desire.

In this perspective, one could critically notice that God is a concept that territorializes knowledge, change, and appetition. It encompasses limitless movement, imposing causal order on infinite dynamics, enclosing the monads in a state of contemplation that binds each soul to itself and confines her influence to local domains. §62 of *Monadology* reveals the narcissism inherent in modern metaphysics:

> Thus, although each created monad represents the whole universe, it represents more distinctly the body which is particularly affected by it, and whose entelechy it is. And because this body expresses the whole universe through the interconnection of all matter in the plenum, the soul also represents the whole universe by representing this body, which belongs to it in a particular way (ibid., p. 27).

Numerous souls and bodies are to be desired and perceived. However, the monad is compelled to commence with its own. Even in Baumgarten, the *fundus animae*

---

[22] We have observed this in relation to the notion of the plasticity of the soul. Regarding the concept of Figure, the fact that it embraces the relationship between the object and the subject is demonstrated (although in a sense that can only remain at a very general level here) by the fact that Alessandra Campo associates this notion with Lacan's concept of the phantasm. See [11, p. 36].

delimits a field that is suddenly appropriated as *corpus meum*, as my body, introducing possession where there is always more and always less than a relation of property (see [5, p. 199]). The aesthetic impetus counters such centripetal motion by propelling the subject towards an exteriority, causing the former to lose its shape and to transform into something else. Aesthetics deterritorializes desire by escaping the narcissistic enclosure of metaphysics, conceiving the *fundus animae* not as an end to be attained, but rather as something playful: the source of a wild, nomadic enjoyment that captures the subject down to the genetic roots of her subjectivity. The plasticity of the soul conveys the following: I can exist within myself, but also in a different form, such as, for instance, the shape of a work of art.

This does not mean that each process of subjectivation is arbitrary. Metastability is not accidentality. If there is no causal necessity, there is at least a virtual alignment of our desiring, "self-molding form", towards which art offers us an unparalleled paradigm. As we approach a work of art (see [27, p. 65]),[23] there is nothing but an anticipation of perception devoid of an object (we do not await the "object" any more than we await one of its properties). The duration of the anticipation does not matter, neither does the number of times the artwork has been viewed in the past, for if our interest is piqued, it will accompany us both as we witness the work of art and as we leave it behind. The aesthetic experience is never complete; it is perpetually interrupted. Perhaps, art may suggest that our reality is always incomplete.[24] The anticipation is a paradoxically aimless orientation. An intensity without source or object. A phantasmatic sensation and a tangible immateriality. Something that is neither virtual, nor sensory in itself.

The main question of this paper—how sensation can anticipate what is exterior to it, namely the work of art—cannot be answered without reference to the appetite of the soul as a perceiving instance. The aesthetic experience demands that, as Herder showed, we think the paradoxical simultaneity of a modal and relational ontology. We cannot become what we are, but rather, always what we are not. Processual becoming presupposes something like this "not" to which we relate a moment before change (modalization, individuation, actualization) takes place. If our question is ever to be exhaustively answered, the enquiry should be posed by beginning with the self-enclosure of the work of art as the asymptotic correlate of an intensive aesthetics.

References

1. Adler, H.: Die Prägnanz des Dunkle. Gnoseologie—Ästhetik–Geschichtsphilosophie bei Johann Gottfried Herder. Felix Meiner, Hamburg (1990)
2. Adler, A.: Théorie de l'art et épistémologie: la Plastique de Herder. In: Décultot, E., Lauer, G. (eds.) Herder und die Künste. Ästhetik, Kunsttheorie, Kunstgeschichte (pp. 193–202). Wolfgang Adam, Universitätsverlag Winter Heidelberg, Heidelberg (2013)

[23] "Let us approach a statue as if enveloped in holy darkness, as if we shall discover there for the first time the *simplest concept* and the *meaning of form*".

[24] A condition that Bataille calls "the insufficiency of being". See [3, pp. 171–173].

3. Bataille, G.: Le Labyrinthe. In: Documents: doctrines, archéologie, beaux-arts, ethnographie (1929) [Trans. The Labyrinth. In id., Visions of Excess. Selected Writings. 1927–1939. Edited and Translated by Allan Stoekl, with Lovitt, C.R., Leslie Jr, D.M. University of Minnesota Press, Minneapolis 1985]
4. Baumgarten, A.G.: Meditationes philosophicae de nonnullis ad poema pertinentibus (1735) [Trans. Reflections on Poetry. Translated by Aschenbrenner, K., Holter, W.B. University of California Press, Berkeley-Los Angeles 1954]
5. Baumgarten, A.G.: *Metaphysica* (1739) [Eng. Trans. Metaphysics. A Critical Translation with Kant's Elucidations, Selected Marks, and Related Materials. Trans. and ed. by Curtney, D. Fugate and John Hymers, Bloomsbury, London-New York 2014; Anast. Repr.: Metaphysica, Editio VII, Georg Olms Verlag, Hildersheim, New York 1982]
6. Baumgarten, A.G.: Aesthetica (1750–1758) [Anast. Repr.: Georg Olms Verlag, Hildersheim, Zurigo, New York 1986]
7. Benjamin, A.: Art's Philosophical work. Rowan & Littlefield, London, New York (2015)
8. Berkeley, G.: An essay towards a new theory of vision. In: Clarke, D.M. (eds.) Philosophical Writings. Cambridge University Press: Cambridge 2008. (1709)
9. Berlot, U.: Duchamp and the notion of optical tactility. In: Alcaraz, M.J., Rubio, M.C.S. (eds.) Art, Emotion and Value Proceedings of the 5th Mediterranean Congress of Aesthetics, 4–8th July 2011 (pp. 471–478) (2011)
10. Burke, E.: A philosophical inquiry into the origin of our ideas of the sublime and beautiful. In: Womersley, D. (ed.) Penguin, New York 1999.
11. Campo, A.: Fantasma e sensazione. Lacan con Kant. Aesthetica Prepint, p. 106 (2017)
12. Colombo, A.: Immanenza e molteplicità. Gilles Deleuze e le matematiche del Novecento. Milano-Udine, Mimesis (2023)
13. D'Angelo, P.: Dal Settecento ad oggi. In: Luigi Russo (ed.), Estetica della Scultura (pp. 91–125) (2023) [Translations by Pucci, G., Di Stefano, E., D'Angelo, P. Aesthetica Edizioni, Palermo]
14. DeLanda, M.: Intensive Science and Virtual Philosophy. Bloomsbury, London, New York (2011)
15. Deleuze, G.: Francis Bacon. La logique de la sensation (1981) [Trans. Francis Bacon. The Logic of Sensation, translated by Smith, D.W. Continuum, London-New York 2003]
16. Deleuze, G. Le pli. Leibniz et le Baroque (1988) [Trans. The Fold. Leibniz and the Baroque, translated by Conley, T. The Athlone Press, London 1993]
17. Duchamp, M.: Notes. Flammarion, Paris (1999)
18. Gaier, U.: Herder's early neoplatonism. In: Noyes, J.K. (ed.) Herder's "Essay on Being". A Translation and Critical Approaches (pp. 162–182) (2018) [Trans. by Noyes, J.K. Camden House, Rochester, New York 2018]
19. Grazioli, E.: Infrasottile. L'arte contemporanea ai limiti. Milano, Postmedia (2018)
20. Halák, J.: Mathematics embodied: Merleau-Ponty on geometry and algebra as fields of motor enaction. Synthese **200**(1), 1–28 (2022)
21. Heinz, M.: Sensualisher Idealismus. Untersuchungen zur Erkenntnistheorie des jungen Herder (1763–1778). Felix Meiner Verlag, Hamburg (1994)
22. Heinz, M., Clairmont, H.: Herder's Epistemology. In: Adler, H., Koepke, W. (eds.). A Companion to the Works of Johann Gottfried Herder. Boydell & Brewer-Camden House, New York (2009)
23. Herder, J.G.: Versuch über das Sein, in Johann Gottfried Herder Werke. In: Gaier, U. (eds.) Band I. Deutscher Klassiker Verlag, Frankfurt 1985 (1763) [Trans. by Noyes, J.K. (ed.) Herder's "Essay on Being". A Translation and Critical Approaches. Translated by Noyes, J.K. Camden House, Rochester, New York]
24. Herder, J.G.: Grundsätze der Philosophie. In: Suphan, B. (ed.) Sämtliche Werke. XXXII. Georg Olms Verlag, Hildeshei, New York (1899) (1766)
25. Herder, J.G.: Kritische Wälder, 4. Wäldchen (1769a) [Trans. Critical Forests: Fourth Groove, On Riedel's Theory of the Beaux Arts, in Id., Selected Writings on Aesthetics. Translated and edited. by Moore, G. Princeton University Press, Princeton-Oxford 2006]

26. Herder, J.G.: Zum Sinn des Gefühls. In: Id., Saggi del primo periodo, edited by Agazzi, E., Gabbiadini, G. Giunti Editore/Bompiani, Firenze-Milano (2023) (1769b)
27. Herder, J.G.: Plastik. Einige Wahrnehmungen über Form und Gestalt aus Pygmalions bildendem Träume (1778a) [Trans. Sculpture. Some Observations on Shape and Form from Pygmalion's Creative Dream. Edited and Translated by Gaiger, J. The University of Chicago Press, Chicago and London 2002]
28. Herder, J.G.: Über das Erkennen und Empfinden der menschlichen Seele (1778b) [Trans. On the Cognition and Sensation of the Human Soul. In: Philosophical Writings Translated and Edited by Forster, M.N. Cambridge University Press, Cambridge 2002]
29. Kaiser, B.M.: On Aesthetics, Aisthetics and Sensation–Reading Baumgarten with Leibniz with Deleuze. Esthetica. Tijdschrift voor Kunst en Filosofie. https://estheticatijdschrift.wp.hum.uu.nl/wp-content/uploads/sites/175/2014/09/1-Esthetica-Onaestheticsaistheticsandsensation---readingBaumgartenwithLeibnizwithDeleuze-2011-10-17.pdf
30. Kant, I.: Der einzig mögliche Beweisgrund zu einer Demonstration des Daseins Gottes (1763) [Trans. The Only Possibile Argument in Support of A Demonstration of The Existence of God. In: Walford, D., Meerbote, R. (eds) Theoretical Phylosophy 1755–1770. Cambridge University Press, Cambridge 1992]
31. Kant, I.: Kritik der Reiner Vernunft (1781) [Trans. Critique of Pure Reason, edited by Guyer, P., Wood, A.W. Cambridge University Press, Cambridge 1998]
32. Kant, I.: Prolegomena zu einer jeden künftigen Metaphysik die als Wissenschaft wird auftreten können (1783) [Trans. Prolegomena to Any Future Metaphysics, edited by Hatfield, G. Cambridge University Press, Cambridge, New York 2004]
33. Kant, I.: Kritik der Urteilskraft (1790) [Trans. Critique of the Power of Judgment, translated by Guyer, P., Matthews, E. Cambridge University Press, Cambridge, New York 2000]
34. Krauss, R.E.: The Optical Unconscious. MIT Press, Cambridge (1993)
35. Largier, N.: The plasticity of the soul: mystical darkness, touch, and aesthetic experience. Modern Language Notes **125**(3), 536–551 (2010)
36. Leibniz, G.W.: Meditationes de Cognitione, Veritate et ideis (1684) [Trans. Meditations on Knowledge, Truth, and Ideas. In: Id., Philosophical Essays, translated by Ariew, R., Gaber, D. Hackette Publishing Company, Indianopolis-Cambridge 1989]
37. Leibniz, G.W.: (1700) Leibniz an die Churfürftin Sophie. In: Gerhardt, C.J. (eds) Die philosophischen Schriften (vol. 7). Georg Olms Verlag, Hildesheim, New York (1978)
38. Leibniz, G.W.: Nouveaux essays sur l'entendement humain (1704) [Trans: New Essays on Human Understanding. Translated and edited by Remnant, P., Bennett, J. Cambridge University Press, Cambridge, New York 1994; fr., Garnier-Fammarion: Paris 1966]
39. Leibniz, G.W.: Principia philosophiae (1714) [Trans. Monadology, translated by Strickland, L. Edinburgh University Press, Edinburgh 2014]
40. Maggiore, V. Il corpo come strumento della sensibilità: Riflessioni sul conoscere e il sentire fra estetica e fisiologia. Studi di estetica **XLIV**(IV), 85–105 (2016)
41. Malabou, C.: L'avenir de Hegel. Plasticité, temporalité, dialectique. VRIN, Parigi (1996) [Trans. The future of Hegel. Plasticity, Temporality and Dialectics. Translated by During, L., Routledge, London, New York 2005]
42. Marelli, F.: Fisica dell'anima. Estetica e antropologia in J. G. Herder. Mimesis, Milano-Udine (2012)
43. Mazzeo, M.: Storia Naturale Della Sinestesia. Dalla questione Molyneux a Jakobson. Marcerata, Quodlibet (2015)
44. Merleau-Ponty, M.: La prose du monde (1950–1952) [Trans. The Prose of the World. Translated by O'Neill, J. Northwestern University Press, Evanston 1973]
45. Plato. Εὐθύφρων (399 BC) [Trans., Euthyphro, in Id., Five Dialogues, Translated by Grube, G.M.A. Revised by Cooper, J.M. Hackett Publishing Company, Indianopolis, Cambridge 2002]
46. Pseudo-Longinus.: Περὶ ὕψους (1 A.D.) [Trans. On the Sublime. Edited by Halliwell, S. Oxford University Press, Oxford, New York, 2022]
47. Sanouillet, M., Peterson, E. (eds.): The Essential Writings of Marcel Duchamp. Thames and Hudson, London (1975)

48. Shepherd, N.: The Living Mountain. Canongate, London 2014 (1977)
49. Simondon, G.: L'Individuation à la lumière des notions de forme et d'information (1958) [Trans. Individuation in Light of Notions of Form and Information. Translated by Adkinds, T. University of Minnesota Press, Minneapolis-London 2020]
50. Tedesco, S.: Studi sull'estetica dell'illuminismo tedesco. Edizioni della Fondazione nazionale Vito Fazio Allmayer, Palermo (1998)
51. Tedesco, S.: L'estetica di Baumgarten. Centro Internazionale Studi di Estetica, Aesthetica Preprint Supplementa (2020)
52. Tuppini, T.: Kant. Sensazione, realtà, intensità. Mimesis, Milano (2005)
53. Wolff, C.: Psychologia rationalis (1734) [Anast. Repr.: Georg Olms Verlag, Hildesheim, New York 1972]

Chapter 4
GENUS

Matteo Bonazzi

> The letter [is an] analogous to a germ cell, which [...] transmits (*véhicule*) life and death together.
> Jacques Lacan.

Abstract The essay explores, through the lens of psychoanalysis and Lacan's thought, the possibility of a "biology of the unconscious" that dismantles the classical idea of the living being as an organized and purposeful substance. By critically reworking the Aristotelian paradigm and rethinking the notions of life and body, it shows how the living does not precede language, but is already its structural effect. In this view, the body is not a foundation but a surface of writing, a site of signifying inscription that produces subjectivity. Life appears not as a natural function, but as the effect of a symbolic cut that exceeds all organicity. When interrogated psychoanalytically, biological knowledge reveals its discursive and structural underpinnings. What emerges is a scene of the living in which life is written, deciphered, and split, rather than merely produced and organized. The text thus outlines a genealogy of human life as a symbolic effect rather than a substantial essence.

4.1 On the Living

4.1.1 What Kind of Biology?

When we refer to biology, we mostly mean the type of scientific research on the living that is imbued with Aristotelian thought. This is what we commonly call *bio-logical* knowledge, because it inscribes empirical research on life and living beings in the categorical *organon* of Aristotelian logic [4].

The discovery of the unconscious had to push beyond the limits of implicit Aristotelianism in our common conception of the living in order to foster a form of

M. Bonazzi (✉)
Università di Verona, Verona, Italy
e-mail: matteo.bonazzi@univr.it

S. Vizzardelli and R. Panattoni (eds.), *Plasticity and Discontinuity*, Lecture Notes in Morphogenesis, https://doi.org/10.1007/978-3-032-03119-8_4

knowledge that could rise to the level of what hysterical patients whispered into Freud's ear (see [3]). In fact, as we know, the discovery of the unconscious had to distance itself from its biological and physiological background in order to follow what psychoanalytically oriented listening gradually revealed to its discoverer.

Why go back to psychoanalysis to reconsider the question of *genus*, generativity, and the coming to life of the living? From a certain perspective, all formations of the unconscious can be seen as more or less failed attempts to grapple with the enigma of life. Consider *little Hans*: what is it that so torments him, enclosed in his phobic representations of horses, their *Wiwimacher*, and the noise they make in the streets of Vienna? As Lacan points out, "(...) this has to do with themes of life and death, of existence and non-existence, and especially of birth, that is, the appearance of what does not yet exist" [13, p. 246].

We may say that the object of psychoanalysis is *a life*, but precisely because it is *one* life, not life in general, psychoanalysis was forced to rethink from the ground up the categories with which we conceptualize life. As we aim to show in these pages, this process led to the emergence of an alternative biological paradigm, one centered on the retroactive effect that speech has on the living. This brings about a fundamental shift, not only in how we think about *bios* but also in how we understand *logos*—that is, how we relate to life, to speech, and to the discourses joining them.

Let us take as a starting point Lacan's so-called "return to Freud." In response to the sense of disorientation he perceived in the clinical and theoretical landscape of post-Freudian thought, the young psychiatrist felt the need to reclaim the core of Freud's discovery, emphasizing its fundamentally anti-naturalistic character. While others took the path of biologizing Freud's findings—consider, for instance, Wilhelm Reich's trajectory—Lacan drew upon the full breadth of contemporary knowledge to return to Freud through a radically anti-biologistic reinterpretation. His key concepts—structure, subject, recognition, desire, imago, and also the limit-concept of *Trieb* (which Lacan insists should not be translated into English as instinct, precisely to underscore its radical non-natural character)—are all situated within an order, first imaginary, then symbolic, which by definition is constituted through a fundamental subversion of the natural and biological realm.

As Lacan emphasized during those years: "Nature, such as it presents to man, such as it coapts to him, is always deeply denatured" (*Idem*). Also, his reference to the Real, which appears from the very beginning of his teaching, carries no biological connotation. On the contrary, it emerges within the perversion of the biological, a process shaped by the imaginary doubling and symbolic inscription, like a supernumerary refuse, that which has been discarded, resulting from the impossibility of thinking and symbolizing one's own death.

This anti-biologistic stance runs throughout Lacan's entire production, rendering psychoanalysis a practice founded on ethical elements that stand far apart from those that inform medicine, which continues to frame care through a biological understanding of the living. This also explains why Lacan, unlike most post-Freudians, assigns particular importance to the death drive and the notion of repetition. Both, in fact, radically overturn biological assumptions, exposing the living being that we are to fixations and death-driven repetitions that are utterly incompatible with the principles of self-preservation and reproduction of the species. Indeed, it is precisely here that psychoanalysis grounds the inadmissible overlap between sexuality and death, which defies all classical biological rationality.

Lacan would continue to distance psychoanalysis from all biological legacies until he succeeded in developing a logic different from Aristotle's, *the logic of the not-all*. Awakening—at least he did, as he ironically remarked—from the Aristotelian dream (see [23, pp. 23–24]) in which we are all caught, perhaps alone, as he had always been in his "relation to the analytic cause" [21, p. 97], Lacan ultimately established another *organon*, one that allowed him to rethink the living in an entirely new way. I view this final return to biology as the "unrealized" germ he left to his son-in-law, Jacques-Alain Miller, and to the "Lacanian biology" [39, pp. 7–59] we will refer to here. The notions of life, the living, generation, and transmission are thus placed within an unprecedented discourse, reinterpreted through the lens of Freud's discoveries in the realm of the unconscious, offering us an alternative way of thinking about biology and engaging with that peculiar living being that happens to speak.

Alongside this remarkable intellectual adventure, it seems to me that this return to biology—*after* or *starting from* Freud and Lacan—may help us confront what the present offers in terms of increasingly extreme experiences that challenge both our capacity to live and our ability to think. Philosophy has attempted to conceptualize these extreme experiences, stretching its/our conceptual arsenal. Also, in light of bio-technological transformations, these experiences are exposing everyday life to a radical paradigm shift.

The implicit biology that informs our discourse on the living is inscribed within the Aristotelian paradigm of becoming whose object is individual substance (*hypokeimenon*). Subjects are conceived as individuals marked by the unity of the body, which comes into being and disintegrates with the beginning and the end of life. The logic of life—what we call biology—is thus embedded within the framework of individual substance. The underlying assumption is that the *one* of the living, which lives and dies, is the individual: the *one* of the body.

However, the extreme experiences to which contemporary medicine exposes us—through the transformations of life's temporalities and modes of being—teach us that nothing is actually this simple. More and more frequently, and unfortunately, we encounter situations where life overflows into death, where what sustains life extends far beyond the individual body, or where what signals life seem to be more in the hands of another who bears witness to it than in those of the individual from whom life appears to have already departed.

It seems we need another biology able to provide us with a different understanding of the living, one that goes beyond the limits and unity of the individual—a biology suited to our times. We need another *one* to think of "life-death," written as a single word, as Jacques Derrida proposed in the 1970s. This paradigm shift requires a prior rewriting of the field of the Other (of knowledge, language, social bonds—in a word, of *logos*), starting from the paradoxical matrix, which however has its logic, provided to us also by how psychoanalysis experiences the status of the object. According to Lacan, in fact, the latter contains the germ cell of life, a minimal and senseless element that reveals the material root of the signifier the human being is made of, its specific causality—that is, one without determination, but precisely for this reason capable of having an effect, as we will see, of "spontaneous generation" [15, p. 297].

The challenge posed by the return of this ancient notion compels us to rethink *bio-logy* (see [38]), and morphogenesis starting from the function that the signifier performs, once it is detached from all forms of linguistic primacy, which is to mark the body and inscribe it into the structure of the Other. Between the body and the signifier, between the living and the speaking being, Freud's discovery of the lost object, which Lacan reinterprets as the object *a*, will serve as our guide in this psychoanalytic adventure revolving around the enigma of *one* life.

4.1.2 Life and Knowledge

Despite everything, we must try to think about the concept of life and the living while knowing, as Lacan himself emphasizes, that we do not have the slightest idea of what life is (see [22, p. 201]).

A first observation could therefore be that life is what we do not know, what lacks in knowledge, what indicates to knowledge that there is always an x that cannot be known. Life, ultimately, could be understood as that which is "not-known."

In this sense, life *uncompletes* knowledge, it paralyzes it, deprives it of its natural tendency toward totality. Life makes *the whole a part greater than itself*—without proportions, incalculable. As we will see, here Lacan's reference is not *his* Spinoza, but Leibniz, when he urges us not to "start from the All, […] it is the part which holds it and contains it" [5, p. 342], [10, 36]. Yet, as Miller points out in his course, it is precisely from these considerations that Lacan strives to conceive of an algorithm of the living—a logic that may allow us not so much to say something about life but to do something with the living being that we are while we speak.

A preliminary consideration must be made here before we continue. In fact, it is necessary to transition from logic understood as a representational structure, aimed at describing, explaining, and cataloging reality, in this case the phenomenon of life, to logic understood as writing, drawing, as the weaving of lines, figures and edges that no longer aim at representation, or seek to speak life, but inhabit it from within, continuously tracing and retracing its boundaries, its emergence, its ripples. In other words, it is necessary to shift from the plane of representation to the plane of immanence.

This means writing life by interrupting its performance, and this is what is at stake in the attempt to recover its algorithm. Writing the real substitutes the clumsy and repeated attempt to represent reality. This writing arises from the failure of this attempt, from the catastrophe of the theater of life, from the encounter with its obscene side. Life begins here, we might say, where it suddenly provokes our horror of knowing, our *not wanting to know anything about it.*

When it appears on the scene, life performs a punctuation. It punctuates the performance with its unique, singular, unrepeatable trait. This is how it occurs, continuously, between one act and another, indicating that, as the expression goes, *life is elsewhere*: exactly because it is always written here, on the reverse side of its continuous performance. It marks discontinuity, rupture, a leap: life is a cut, a fissure, a tear. And this is how it always escapes knowledge—which either anticipates it too soon or grasps it too late. As Lacan wrote, life is the "mis-taking" (*méprise*) [26, pp. 329–339] of the subject supposed to know: the point at which the grasp (*greifen*) of the concept (*Begriff*) drops what matters most, because the more it tries to seize it, to tighten its grip, the more knowledge lets go of life. So, life is nothing but this continuous loss, this knowledge-in-loss, this grasping without a hold: *méprise*.

In attempting to say all this, we are beginning to outline what is at stake: what Lacan called enjoyment, whose logic is modeled on what Freud discovered in his clinical work and named the "lost object." And it is precisely here, at the intersection of the three registers of experience, that Lacan in-scribes what we will call the *point of life*.

The first consequence is that life does not flow, it does not extend outwards, it has nothing to do with the broad plane of relationships, contexts, or fields of action. Life is always written precisely at the same point, intensively, and at the same time, in a non-negative way. *That one enjoys* continues to be written behind what is said, precisely at the point where silence also occurs. To say that life is always elsewhere does not mean operating according to a kind of negative theology of the living, one that always eludes or refers to something else. Rather, it means indicating that life is what escapes the here and now, what erases itself at the very moment it is inscribed. Japanese calligraphers know this well when they cover city squares with magnificent ideograms painted in water-based ink that evaporates, step by step, in the light and heat of the sun. If life is enjoyment, it is so precisely insofar as it is lost at every step—and it is precisely because of this that it allows the step, our step, to move forward.

We must, therefore, begin to remove life from representation, to remove *one* life from representation. This is the function of the algorithm according to Lacan: to stop speaking about life in order to begin inhabiting it. In other words, to write it. Formalization reduces representation almost to zero. And it is here—when we no longer know anything about it, when supposed knowledge collapses—that life moves towards us. When we stop chasing after it, because it is always elsewhere, it is then that life finds us. Lacan's algorithm of the living makes it possible to swap planes: we go from planning to stumbling, we go from a life that is anticipated, projected forward, to one that happens and moves toward *us*.

At the same time, however, life must also be removed from full formalization. The algorithm must fail for life to happen. It is necessary to avoid the situation in which life, once removed from imaginary representation, is entirely inscribed within the signifying machine, which would turn it into an endless metonymic displacement, trapping us in the dimension of what is known as bad infinity. No, life belongs neither to the order of the Imaginary One, nor to the Symbolic Two of differential signifying displacement. *Another One* is needed to think the *point of life*, and this is where Lacanian biology makes its contribution.

4.1.3 From One to Another

The One of life take shape as the unity of the individual body. This is the implicit starting point of all biology: life is the life of an individual who is born and, for this reason, also dies—an individual whose living unity derives from the body-one it bears. "My" life is, first and foremost, the life of "my" body: it depends on it, not on me—or at most, on how I use it. In any case, life and death can be discerned based on the unity of the body itself, which indicates its life and/or its death, with all the complexity this entails, given that bodies today can extend their individual unity thanks to machines, prostheses, and biomedical supports, which serve precisely to preserve life. What lives is the body in its individuality.

The notion of enjoyment and that of the lost object led psychoanalysis to emancipate itself from the identification of life with the uniqueness of one's own body. Enjoyment does not seem to follow the logic of self-preservation of the individual body. It seems to push beyond life, forcing those who listen to it speak to rethink their own categories. This "beyond" was initially understood by Freud as "before"—a regressive drive to return to the beginning: to before birth, a place marked by the nostalgia of an originary melancholic call. Lacan's indication, however, was different. By introducing the concept of enjoyment in place of the death drive, he sought to free life from the biological dimension that had implicitly shaped it. From the point of view of the biology of the individual living being, enjoyment would thus appear to be a return to the inorganic. But if we let go of the framework of representation through which we habitually think of life in terms of the individuality of the body, then we discover—to our surprise—that enjoyment is not what is missing, it is not a nostalgic, melancholic return, but rather an excess, a surplus of life. Enjoyment tends toward another life, *a life beyond life*. And this surplus of life can be grasped only on the condition that we abandon the One of the individual body, to which traditional biology remains anchored, in order to move toward the One of the signifier. Lacan's thesis, Miller states, is "that the One comes to us from the signifier and not from the One of the body" (Miller 2000a).

We have made progress and found a new *one*, but we are immediately faced with a new risk: that of holism, which we find to a large extent also in contemporary immanentism. If life is One, it does not follow that *all is One*: in fact, "The Great Pan is dead" [15, p. 198]. Rather, Lacan's Spinozism leads us to say that "God is unconscious" (Ibidem, p. 59). Thus, life that insists, life that pushes beyond itself, is *one* without being all. It is the One of partiality—without field, without background, without totality: the One of singularity.

In this sense, enjoyment must be conceived as starting from the One of the signifier, but at the same time as that which is erased beneath each signifier. The One is lost, but not denied: the One always occurs in the form of two—as *more* enjoyment of what is *less*.

4.1.4 Deconstructed Bodies

The early Lacan returns to Freud through this insistence on the difference between life and body. The life in question, starting from Freud's discovery of the unconscious, has to do with death: it arises from the understanding that the point of its fullest realization in also the extreme passage to the impossible of death. This is why sexuality is the guiding compass: the discovery, that is, that "language does not dominate—from this foundation of sex in so far as it is perhaps most profoundly linked to the essence of death" [18]. First and foremost, therefore, it is necessary to deconstruct the biological identification of life with the individual body, which is precisely what Lacan's early account of the mirror stage contributes to (see [30, pp. 75–81]).

The fact that the individual body is a sign of life means we attribute something real to the body. However, Lacan initially holds that we have nothing of the body except its image. This image is certainly decisive, at least in its ability to unify the "fragmented body" (Ivi, p. 70) we are at birth. For this reason, upon closer examination, what is born is not at all the individuality of a living body—this individuality exists only for those who witness the birth, namely, the adult who anticipates it with his or her own desire. The birth of the body as *One* requires a whole series of successive conditions that reveal the merely constructed dimension of the individual body, upon whose assumption rests the implicit biological discourse that informs our way of thinking about subjectivation as a whole. As Lacan would recall in Nice in 1974, many years after his first public presentation of the *mirror stage* (in 1936):

> Man loves his image like that which is most close to him, that is to say, his body. Simply put, he has strictly no idea what his body is. He believes it is 'me'. Each of us believes it is oneself. But it is a hole. And, then, outside of it, there is the image. And with this image, he makes the world [35].

The imago of one's own body forms a one, and thus it generates the illusion that this is where life resides—in the unity of the body, the real place of the living being. However, this is an illusion, an impression caused by an excess of love, the love for our image, for our body, linked to the fact that we believe it to be 'I'. It is indeed the ego—though as Lacan and Rimbaud remind us, "I is another." The subject is not there; he does not live with his body, just as Hamlet says of the king—and Lacan suggests we "replace the word 'king' by the word 'phallus'" [14, p. 353]: "*The body is with the king, but the king is not with the body.*"

Life is elsewhere and unfolds on another stage—the stage of the unconscious. Here, however, we discover with horror that we know absolutely nothing about the body, because, in reality, "it is a hole." This is the real "hole" that Lacanian biology is called upon to think. Meanwhile, we continue to think of ourselves in images, constructing mirror images of ourselves, narcissistically, so as to make the world and reality a perfect double—a reflection in which we project the joys and sorrows of a life that does not exist precisely because it is imaginary.

4.1.5 The Second Life

There is, then, a second life—one that is written on another stage, that of the signifier. Here, the One is not the body but the signifier. The phallus, as a symbol of life, is this signifying element that inscribes us in the field of the Other and delivers us to the intermittent pulsation of life-death, embodying the minimal rhythm of *fort:da*. This alternation both grants and takes away life: it makes the being of the body fall away, dissolving our identification with the imago of one's own body, and replaces it with the body that is distanced in the register of 'having'. I *am* no longer my individual body, I *have* a body. This introduces a fundamental division between subject and body, which, in effect, is the first (and last) Other one encounters. The subject therefore loses the body that it is and gains another one, the body it has. By having a body the subject must relate to the alterity which inhabits it in its most intimate core. *My* body is always Other, and it is only like this that I can have it—no other way. Yet, this loss is also a gain: less life—which delivers the being (of the body) to the nostalgic longing that so weighed upon Freud's conception of the drive—is also more life, unleashed by our having (a body), precisely because in the dimension of having we inhabit the intermittence of presence and absence brought about by the signifier.

The One here is no longer the one imaginatively signaled to the individual by the imago of one's own body: *I am my body*. The One is thus radically de-substantialized, being reduced to the threshold that marks the passage from one to the other, from presence to absence. Jean-Luc Nancy writes that "We lose our footing at the body" [40, p. 13]. There, in the midst, between presence and absence, the body I have occurs, at the point of falling. One thus gains access to the life of a body that *occurs* precisely because it *is not*, which therefore continuously transits in the pulsation that signals its own living presence. We move from the 'I', which is another, to the subject, which is the Other of this signifying pulsation. It is/it is not: this is the passage from being to having (or not having).

This fundamental paradigm shift allows Freudian psychoanalysis to reconsider biology in a new light. This new biology, however, no longer conceives of the living being within the narcissistic bubble of the imaginary. It exposes the living being to a beyond, where life spills over into death and, in doing so, signals its being alive. Because the living being is reconsidered from the perspective of this signifying "denaturation," bio-logy is now taken literally: there is life because there is logos, signifying writing, which brings it into existence as such. Which does not mean it does not constantly expose it to death, for in the logic of the signifier "life-death" is inscribed from the cut that the signifier imprints upon the body of the living being, thereby making it exist. In this paradigm shift, if life is no longer what the (individual) being signals in the mirror, death too changes its status: it is no longer so-called biological death, the death of the living being that, as Heidegger would say, simply perishes. Rather, it is a second death, a symbolic death, brought about by the beyond that the signifier opens in the speaking body. Once symbolized, the living body is no longer simply the individual body, which either is or is not, but the *speaking body* (*Sprachleib*), always exposed, at every moment, to the elsewhere to which it is destined to by the signifying rupture. It is another life and another death: the living being marked by the signifier is exposed to an *excess of life* that the symptom signals as an "event of the body" [27, p. 569].

Here the field of Lacanian biology opens up. Here we no longer have the law of adaptation to the environment or, psychologically, the pursuit of well-being, but rather the law of enjoyment *beyond*, enjoyment that manifests itself in the places and forms of repetition. Instead of the hedonistic reason that guides biological adaptation, we have drive satisfaction, which continuously flirts with death. As Miller points out:

> Lacan's complex exertion touches on both death and libido. It consists in showing that death is by no means the prerogative of the death drive, that it is present in the sexual drives and, symmetrically, the libido is present in the death drive. (…) The essential Freudian dichotomy is re-absorbed somehow by Lacan who evinces that death and the libido have close links. This is the real sense of his myth of the lamella: the libido is a deadly being [38].

Instead of evolutionary continuity, we thus encounter the radical discontinuity brought about by the signifier. Its marking upon the body exposes the living being to a double creationism: on the one hand, the signifier emerges from nothing, like a cry in the night that retroactively brings into existence the silence it interrupts; on the other, it is only through the occurrence of the signifier that nothingness itself is generated as its destination. "*The Ethics of Psychoanalysis*, in this regard, aims at nothingness in its double aspect: the nothingness that arises because of the signifier and the nothingness that brings forth the signifier" (Ibidem). In between, between one signifier and another, lies the void of life, which, in repetition, causes the senseless pulsation of the drive. This is what life becomes at this level of discourse: it is in the emptiness of this nothingness that the echo of the drive resonates, in the superegoic dimension as well as in the causative thrust of enjoyment.

4.1.6 Dehiscence

In its structural diplopia, the signifier continuously shifts from its dialectical role, through which it traditionally serves to kill the Thing, to its immanent opposite, whereby it becomes a cause of enjoyment. Through this dual movement, it opens the body like a fruit to a kind of "spontaneous generation."

> [*Allow me*] to invite you quite simply to think of Van Gogh's old shoes—on the basis of which Heidegger has given us a dazzling image of what a work of beauty is. You must imagine [the] clodhoppers (…) *ohne Begriff* (…) if you are to begin to see Van Gogh's own clodhoppers come alive with their own incommensurable quality of beauty. They are simply there; they communicate a sign of understanding that is situated precisely at equal distance from the power of the imagination and that of the signifier. This signifier is not even a signifier of walking, of fatigue, or of anything else, such as passion or human warmth. It is

> just a signifier of that which is signified by a pair of abandoned clodhoppers, namely, both a presence and a pure absence—something that is, if one likes, inert, available to everyone, but something that seen from certain sides, in spite of its dumbness, speaks. It is an impression that appears as a function of the organic or, in a word, of waste, since it evokes the beginning of spontaneous generation [15, p. 297].

If the signifier, as conceived by linguistics within a logic of deferral and metonymic slippage, "kills the Thing," the letter—that is, the signifier that falls outside the chain and, instead of referring to other signifiers, interrupts signification—persists in repetition, producing a vitalizing effect. To fully grasp this transition, one need only consider the status of ruins (see [7, pp. 101–115]). These are the remnants of cultural lives that have faded from history, one might say, that have passed on to a better life: stones, fragments of a bygone era, whose traces mark the eternity of a life frozen in time because it has been inscribed into inorganic matter. The analytical experience is shaped by this mode of transmission and temporality—not by subjective time, which psychologically insists on questioning the success or outcome of the cure, but by geological time, as Fedida puts it. Ruins are written through stones that, as Heidegger emphasizes, are without a world, but precisely for this reason, signal the rhythm and modes of a transmission that remains entirely imperceptible—just as it is with the unconscious.

Spontaneous generation thus occurs through ruins that allow the "refuse [*rebut*] of the aforesaid (humanity)" to pass without any apparent transmission [24]. The imperceptible that takes place here is inscribed on the reverse side of symbolic transmission and on what its tombstones attempt to preserve. Rather, it is a passage marked by pure loss, which allows the transit of what has never been. In this way, the object passes from hand to hand, no longer through the Hegelian conceptual recollection, but via a repeated *mis-taking* that establishes a new kind of community around the empty center of this impossible and *immemorial* passage (see [2, pp. 333-342], [1]).

If the sign indicates the individual's presence, the signifier wounds its flesh, definitively de-substantializing its presence. This is how it consigns living beings to the rhythm and alternation of presence and absence. Living presence is therefore this coming and going of life and death, which, in the imperceptibility of the object, gives consistency to its happening here and now.

Ruins allow us to take a further step. It is no longer the *fort:da* that the signifier has impressed upon the body, but its refuse. Not what remains in the sense of that which is continuously lost, which humanity tries to hold onto, but what escapes sideways from the very logic of the signifier and its symbolic grasp. Ruins suggest an apparent return to the sign, to a presence that lingers without going beyond. In this sense they are ruins of a world that reveals a nonworld: it is here—without image and without symbolic representation, as Lacan says of Van Gogh's peasant shoes—that the field of spontaneous generation opens up.

Between the subject and substance, between Hegel and Aristotle, Lacan situates the alternative of the *parlêtre* (speaking-being), the logic of which is illustrated by ruins. As Miller emphasizes, following Lacan: "The *parlêtre* is the union between Aristotle's *hypokeimenon* and *ousia*—the union of subject and substance, signifier and body" [38]. The refuse of the subject supposed to know, its *mis-take*, what is written to the side, has no substance; it is no longer the presence of a being, nor is it a signifying reference, as the subject, by definition, manifests itself in being nothing other than what one signifier represents for another. *It neither stays nor shifts*: it lingers. It speaks, not to say, but to write the "event of the body", the event that is written between the signifier and the body, between the subject and substance. This *between* is decisive for rethinking biology in light of psychoanalysis.

4.1.7 Another Life

The signifier kills the Thing, and the subject-effect that is generated is, by definition, mortifying. "Lacan's barred subject writes the subject—but only insofar as he is already dead. [...] What Lacan calls the subject remains entirely outside the body, outside of life" (Ibidem). Life is elsewhere: each of us interprets it this way, imagining other possible lives, horizons, worlds different from the one mortified by the signifier that embodies it. The lives of others thus exert their fascination on each of us, even when those others are ourselves in different phases and configurations of our existence. This happens, for example, when we see ourselves in old photographs or see our loved ones photographed in our absence (see [6]). The bite of the signifier, which erases us in order to make us exist as subjects, also makes us dream.

Rather, it is about lingering, about learning *to dream more truly*, as Nietzsche pointed out (see [44]). Not elsewhere, but here, where life is written sideways, both mortified and enlivened by the signifier. The traumatic point is the root of this mortification that is constitutive of subjectivation, but at the same time, it is the incision of the fundamental affect that singularizes experience. On one side there is no relationship, being is lost forever; on the other, precisely for this reason, there is love, there is encounter, there is singularity. A *surplus* of life is written on the catastrophic reverse of birth and its trauma. The wound that remains speaks of a real that distances itself from its cynical turn precisely because, in the radicality of this non-relationship, it establishes the possibility of spontaneous generation.

The paradigm of this other life, no longer dreamed but embodied, is, as Jacques-Alain Miller emphasizes, the *Schreberian body* (see [12]). Here, the discussion becomes particularly delicate. In Schreber, the imago of one's own body could not stabilize due to the *foreclosure* of the paternal signifier. Instead, as Lacan points out, the eroticized image of his body takes on a compensatory function. The onset of delusion, therefore, has a stabilizing role and allows for a compensation of the paternal lack. However, one step before delusion, we find the logic of the living that interests us: "What Schreber experiences is that which, through the living body, goes precisely beyond the symbolic order—beyond the signifier" [38]. It is a body that escapes the grip of the symbolic and that also remains as a residue beyond the delusional construction of the imaginary: another life here becomes spontaneously generative. Instead of the phallic One, which writes the body, here we need another One in order to linger. This is how, in the Borromean knot, at the center, where the three registers are joined, in place of the Name-of-the-Father, Lacan writes the object *a*: that is, *life, about which we know nothing*. This not knowing anything means we are all relatives of Schreber, that we must reach that point in order to realize that we know nothing about what matters most to us. At this abysmal point it is about stopping just before the onset of imaginary delusion. Here a desert opens up, the *un-being*. Its "debris" forms One. No longer the phallic One around which the master's and the hysteric's signifying logic revolves, but the object that serves as a compass for the ethics of psychoanalysis.

4.1.8 Corporization

Lacanian biology thus indicates an inverted writing of the living. And it raises new questions for the ancient psychoanalytic dream of rethinking biology based on the principles developed by psychoanalytic clinical work.

This dream was already pursued by the pioneers of psychoanalysis, particularly by Ferenczi in his *Thalassa* [10]. In it, there is a continuous going back and forth from biology to psychoanalysis and vice versa, which unfolds around the enigma of genitality. I would say the work touches upon the very issue we are investigating, using different words and a different conceptual framework, namely generation, transmission, *genus*. It is from coitus that the Hungarian psychoanalyst extracts the logic of the living, based on a synchronic and intrinsically contradictory alternation. As always in psychoanalysis, we start from the symptom. The trajectory of the symptom is the gateway to the unconscious. In Ferenczi's case, this involves "genital stuttering" (Ivi, p. 9)—ejaculatory disturbances that allow him to address genitality. Stumbling, limping, a break in signification: the intermittent structure of the unconscious comes into play right at the beginning. Sexuality thus loses its special status and becomes a generic writing of the drive-based zoning of the body. Beyond and before the primacy of the genital, *genus* indicates this pulsation, this counterpoint that outlines, step by step, the autoplastic (through sleep) and alloplastic (through coitus) topography of the sexed body. The structure of repression, discovered by Freud, is found anew within the biological fabric that informs the living, between the urethral instinct to release and anal retention, giving rhythm and differential articulation to the body, which thus writes the diffusion of libido and draws its zoning.

Scratching, dropping, rubbing, detachment, separation, friction, preservation, displacement, retention: these are all ways of naming the action that traces the contour of the drive object in the body marked by the signifier. Ferenczi proposes a biology of pleasure, in which anal and urethral drives in the process of ejaculation give substance to Freud's concepts of *displacement* and *condensation* (Ivi, p.82). The liminal concept Freud invented with the term drive becomes the key to reading another way of thinking about the living and its biology—not one centered on a utilitarian reduction serving development and survival, but one capable of embracing the paradoxical entanglement of life and death inherent in the functioning of drives. At its core, no longer do we find its symbolic transfiguration on the dialectical stage of representation, but rather the compulsion to repeat the pathogenic and traumatic core that marked the living being at birth. This is its pulse, its character, its stroke. Against the backdrop of the great intrauterine "catastrophe"—the repetition of the primeval phylogenetic aquatic existence—*genus* writes both inscription and belonging, and *exscription*, proliferation. In response to the drive toward "thalassal regression"—*das Ding*—the living being is called to draw the separative object that orients its libido. To transition from the *oceanic feeling* to what we might call a *Mediterranean* navigation: an archipelago of partial objects enabling a certain metastable balance between sea and land. It is in this navigation that the cartography of the senses is written, rendering smell, as Ferenczi says, "the biological prototype of thought" (Ivi, p. 71).

But how can we turn this Freudian discovery, which overturns our notion of biology and the living, into something other than a myth? How can we distance the drive from its mythological representation? This indeed was Lacan's objective, when he showed us that the paradox at the core of human existence, this intrinsic contradiction of sexuality—as Alenka Zupančič has brilliantly highlighted in her book *What Is Sex?* [46]—undermines all ontology, exposing us to the theme of the unrealized, neither being nor non-being, and ultimately depends on the logic of the signifier. Once demythologized, the drive becomes a signifying effect/affect. The fundamental affect of the living is inscribed alongside the signifier that carves into it, and in doing so, keeps it alive. For this reason, the logic of the living depends on the contradiction of drives, life-death, as an effect of the logic of the signifier. The living being depends on this fundamental pre-ontological affect, which forces us to go beyond structure.

Beneath structure, we find the logic of the living affected by the signifier. There is no dialectic, but bio-logy, a writing between life and death of the living being, starting from the fundamental affect that does not manifest at the structural level but only, as we find in *Seminar XVI*, at the "substructural" level [42, p. 3]. Two years earlier, in *Seminar XIV*, we already encountered a revision of the classical theme of alienation, in a non-dialectical form. While in the classical temporality of structure, alienation takes on an imaginary or symbolic form, at the substructural level, we discover that alienation has a real dimension. How can this be expressed if not through a topological twist that language struggles to translate? The speaking body extracts from the Other a minimal and insignificant element, neither imaginary nor symbolic, which it *aspirates* (see [8]) into itself, making it the cause of its own desire. Alienation is real in that it neither dissolves into the Other, as in the imaginary, nor identifies with the Other, as in the symbolic, but instead topologically encircles the hollow, the *objet a*, in the place of the Other in such a way as to extract it and aspirate it into itself as an *alien* cause of its own existence. This is the object in which the subject is "internally excluded" (see [31]), as Lacan puts it. This fundamental affect is the point of the real that is not without the Other. It is the point of life (-death) that keeps one alive. It is the point of repetition that *constitutes the subject as happy-go-lucky* (*heureux*) [21, p. 22] and, in doing so, also makes it generative, suddenly revealing that psychoanalysis, instead of posing as anthropology, actually opens up something that could be called a new history of "nature" (see [9]).

4.2 On the Speaker

In his *Three Essays on the Theory of Sexuality* [11] there is a key term Freud uses when referring to puberty: *Umgestaltung*, translated as "transformation" in English and "*metamorfosis*" in Spanish. Unlike dialectics, metamorphosis does not operate according to the primacy of time, *telos*, or in view of a gain in meaning. Metamorphosis is a logic of the living that primarily involves space, the plastic transformation that inscribes itself between its surfaces, the rhythm of variations, and above all, the possibility that different forms and moments can coexist, constituting complex fabrics with stratified functions.

This expression, used by Freud, reminds us that a significant part of his reflection resists a dialectical translation and speaks instead of becoming, of a morphogenetic process that writes the transformation of drives, the distress of which, for example adolescent distress, would be the symptomatic appearance (see [43, pp. 72–92]). In this sense we could speak of a dialectical Freud, interested in the dynamics of identification that characterize subjective life, and an immanentist Freud, focused on the primacy of the object in its unconscious dramatization. It is starting from this *other* Freud that we can rethink subjectivation in biological terms.

If we consider the present situation, this duality seems to be still in place. Also today we might say that there are two ways of considering subjectivation, one is Foucauldian, the other is Lacanian (see [41]). In both cases, at the center of the process of subjectivation, we find the commandment of knowledge. The fact that knowledge is not an attribute of the knowing subject but that by which the subject is substantively constituted is the defining feature of our time—this view of knowledge was inaugurated by Hegel with his well-known assertion that truth is not substance but subject. The *imperium* of knowledge dominates subjectivation also in its seemingly liberated form, which is self-knowledge. In this sense, knowledge is evidently, for both Foucault and Lacan, always power exercised over the subject, thereby producing its ongoing subjectivation. However, the two approaches differ with regard to a third term: enjoyment. The centrality that our epoch has assigned to knowledge not as a means but as the very substance of subjectivation—we may think of Lacan's notion of the "subject supposed to know" (a synonym for subjectivity)—offers us two different articulations of the Superego and two distinct ways of inhabiting the enjoyment that sustains it. The first is the classical one, the genealogy of which was provided to us by Foucault. The second is the one that led Lacan—not without difficulty and following a series of shifts—to formulate his version of "knowledge-power" as "knowledge-enjoyment" [19, pp. 255–267]. We will therefore begin with the first, reread in light of the insights provided by the second.

4.2.1 Knowledge-Power

The relationship between knowledge and the subject is not a linear one. Any deterministic causality is complicated by the topology that characterizes their double implication. Lacan refers to this relation without relationship with the syntagm "not-without": the subject *is not without knowledge*. This means that there is an intimate connection between the two, but at the same time, a continuous generation of difference, a disjunction. This is because the subject, as we emphasized, is not substance but process, one that however unfolds within a structure of knowledge-power, which constitutes its fundamental grammar.

Between knowledge and subjectivation, a continuously shifting metastable equilibrium is written, indicating the structure of lack that characterizes both terms of the relationship. It is precisely because of this lack that something can occur between structure and subjectivation: if we were to reduce the structure to the order of being or subjectivation to that of becoming, neither would exist. Subjectivation can, on the contrary, take place at the point of structural incompleteness of knowledge, activating its own morphogenesis, which is always conceived in relation to a law or a principle capable of orienting it.

The two fundamental characteristics of subjectivation are therefore lack and the Ideal. A triadic schema is established through the effect of subjectivation produced by knowledge-power, made possible by the lack that constitutes it and is directed by the Ideal. Implicitly, this is a theological/teleological schema, in which lack creates space within the being of the structure for subjective becoming, and the Ideal orients its metastability toward an end. Subjectivation, therefore, exists only where there is both lack and orientation—namely, within the symbolic register. In the imaginary and the real, by contrast, nothing is lacking.

Now, the crucial point is that psychoanalysis has come to understand the failure of this triadic schema, its fundamental crisis, the collapse of the Ideal that once worked in conjunction with lack. The result is a certain retreat—either into imaginary becoming or into the terrifying static nature of the real. The dissolution of the father, in this sense, is not merely a historical event of sociocultural significance but an epochal shift that gave rise to the very adventure of psychoanalysis. And it is precisely because Freud, by listening to neurotics, was able to perceive this crisis unfolding that the triadic structures governing subjectivation became clearer. In other words, with the discovery of the Oedipus complex, Freud was already bearing witness to its obsolescence and this is why he was able to provide the logical interpretation of it, which he passed down to us.

But where does psychoanalysis position itself, and the psychoanalyst, in order to continue speaking of and using the structure and function of the Oedipus complex, despite what we have just said? In the place where, as Lacan emphasizes, the knowledge-power nexus—centered on lack and the Ideal—enters into an irreversible crisis and reveals its radical disjunction. Which means that the disagreement, now structural, between knowledge and power thus requires a different grammar to be understood. The affirmatively contradictory movements of dialectics are not enough;

the dynamic incompleteness of structures is not enough. A new topology is needed, one we will call "substructural" (see [45]). It is Lacan who elaborated it, highlighting that what is at stake in the relationship between knowledge and power is not only lack and the consequent orientation, but also the new fissure that opens up between them. The ethical task of psychoanalysis is therefore to invite us to name, case by case, this structural disjunction which, in other terms, we may define as "the contemporary."

According to Lacan, Freud, and before him Marx, would have observed that in our time, knowledge and power can no longer coexist due to the collapse of the function once guaranteed by the Ideal. The effect of this separation does not produce a new *Weltanschauung* capable of absorbing the crack within itself. Nothing can fill this gap, which psychoanalysis invites us to recognize, first and foremost, for what it is: trauma.

The divergence between knowledge and power manifests itself in symptoms—points of economic concretion that, on both an individual and collective level, signal the crisis of any possible cosmology and, with it, of any general theory capable of representing life and the living. Waste, luxury, the disjunction between production and distribution—all of these are symptoms that, both in the individual and in society, reveal the presence of a *surplus* (*surplus value/ surplus enjoyment*) in place of the *less* that, as we have seen in its trinitarian configuration, the structure was fundamentally premised upon: this excess returns as *less* at the level of distribution. No ideal of justice can any longer guarantee the promise of a fair distribution of wealth or value; nothing resembling the highest good can ethically orient individual satisfaction.

The history of psychoanalysis can be understood as a long and slow attempt to resist this discovery—what Lacan would later call the structural "non-relationship." This resistance is evident in post-Freudian theories centered on object relations, which sought in every possible way to seal the fissure that Freud's listening detected, relying on homeostatic principles of reality that replaced the lost Ideal with the object being promised by an evocative cure. It is precisely here that Lacan, on the contrary, extracted the radical nature of his *objet a*, which is neither objectifiable nor subjectifiable, except in the rigorously topological manner of an "internal exclusion." Faced with the impossibility of tying the trinitarian knot again, Lacan's object indicates that "we are all patients" [19, p. 259]. As subjects exposed to this radical disjunction between knowledge and power, we are all subject to the discomfort of civilization, which writes in each one of us the symptomatic invention that is uniquely ours.

4.2.2 The Empire of Signs/senses

"In the beginning of analytic practice was, let us recall, love" [16, p. 4]. This axiom of Lacan is often interpreted in terms of the phenomenon we commonly associate with transference. What is often forgotten, however, is that at its origin, transference is not a phenomenon but a signifier that impacts the living being, "crystallizing" and dragging it into the chain of signification, into the structure:

> For a crystal to form from a solution described as metastable, the presence of a crystalline seed is necessary. It is from the imbalance induced by this local singularity that a chain reaction begins, ultimately converting the entire solution into a crystal. Similarly, we can say that the symptom "crystallizes" starting from the signifier of transference [37].

This is how we must understand Lacan's assertion, which we placed at the beginning: "the letter is the germ cell of life." In the sense that, by crystallizing, the signifier breaks the chain of signification and grafts itself onto the speaking body, producing a solution that progressively radiates outward in what we then commonly call the phenomenon of transference.

It is precisely this point of impact that Roland Barthes sought to remind us of when, at the *Collège de France* in 1977, to the astonishment of his audience, he opened his course on Literary Semiotics by declaring that "Language is fascist not because it prevents speech but because, on the contrary, it forces speech." At the origin of transference, then, lies command—the S_1 that innervates the living being, inscribing it into the chains of language. At the origin is the "You must!" of the Superego—the Empire.

For the master signifier, as Lacan calls it, to be able to grasp something of the living there must be a simultaneous double capture: on the one hand, as Freud had already noted, the empire performs the function of the Ideal that imposes its Law on the living by inscribing it into the order of civilization; on the other hand, as Lacan highlights, the Law alone is not enough. This is why the empire is always also an "emporium" [19, p. 256], a place where things are bought and sold, where not only money, use value, and exchange value circulate, but also enjoyment. The "you must!" of the transferential commandment, which sets in motion the humanization of the living begins, is simultaneously a categorical imperative and a drive toward enjoyment, as Lacan emphasizes in his writing *Kant with Sade* [32]. The origin is the traumatic grasp of the signifier on the body, which turns it into a speaking body subjected to the *empire of signs*, as Barthes ([20], trans. 1983) wrote in the book that bears the same title. But also, and this is Lacan's novelty, the speaker is subjected to the *realm of the senses*, to quote the title of a well-known film by Nagisa Ōshima (*L'Empire des sens* in French) that came out in the same years as the publication of Barthes's book. It is a double empire: the empire of signs and the empire of the senses.

There is no inscription in the structure, and therefore no beginning of life, without this diplopia of the signifier, which is at the same time a dialectical grasp, operating in the sense of *Aufhebung* (from simple signification of the individual to the humanization of the living in general), and an incorporation, incision, writing, production of enjoyment.

It is at this level that Lacan takes us when he corrects his first statement that *the signifier kills the Thing* [33, p. 319]—undoubtedly true but not real enough to explain the functioning of the signifier's grasp in its substructural dimension—with his second statement that "The signifier is what brings jouissance to a halt" [25, p. 24]. The empire kills, elevates, and produces, all at the same time. And this "at the same time," which precedes all temporal articulation, is both an instant and an empty extension.

What interests us here is the point of inscription, the event by which the living is inscribed in the chain of the signifying structure. This is the secret of psychoanalysis: the fact that the Other does not exist (see [14]), and yet, precisely because of this, the insistence on the repetition of what will provide a semblance to this evanescent and metonymically reiterated subsistence.

Let us attempt to formalize what has been outlined so far. The signifying chain, as structural linguistics teaches us, involves the metonymic slippage from one signifier to another. No signifier exists in isolation; by structural law, each signifier refers to another signifier, and so on, indefinitely: S_1-S_n. This is the machine of signification sustained by the empire of signs. And this is the machine that fails, that jams, and that in doing so, provides psychoanalysis with its secret. Which consists in the fact that what the subject seeks, going from one signifier to another, is precisely what was lost at the origin, when the subject was captured by the empire and its double functioning: moral law and absolute enjoyment. This empire, its command, in fact, takes hold of the living because it exposes it to a limit that simultaneously curbs its infinite enjoyment and confronts it with a beyond—that of the father of the horde—whose defining characteristic is not only that he is not subject to the risk of life, that is, death, but also that he can do anything, meaning he is the bearer of absolute enjoyment. The principle of the empire is beyond the law, beyond castration, and at the same time, the emblem of absolute enjoyment.

Here I would like to report two dreams by a young patient, about whom I will say nothing, except that there is an obsessive trait.

In the first dream, he finds himself on a cruise ship, in the middle of the sea, and suddenly discovers he is immersed up to his neck in a container—a foreign object, almost never present on a cruise ship, as he will later note. And yet, this is precisely where our subject has ended up, in this obscene, incongruous, and alienating object. But the worst is yet to come. The container is actually a deposit for excrement, a sort of large sewer in which the subject is about to drown. He has taken refuge in the most incongruous object, the one that is most alien to the scene the dream dramatizes. He is *internally excluded*, then, in a double sense: the subject has ended up in the object that excludes him from the scene; but at the same time, he is excluded internally, from inside the object. In fact, he does everything he can to get out of it and, finally, manages to save himself by dreaming of vomiting. It is at that moment that he wakes up, startled: the disgusting and nauseating object he had ended up in somehow passes from *outside* to *inside*. The struggle to get out of it fails, and so, having absorbed into himself the object he was defending himself from, he can finally expel it. *Austossung*, says Freud, is the act by which the child affirms itself via *negationis*. Subjectivation begins this way—by absorbing the object in order to ultimately spit it out. The outside must pass through the inside to truly be outside, and only then can one meaningfully speak of an inside.

In the second dream, which follows shortly after the first one in the subject's conscious temporality, we no longer have the object but the Thing. In fact, the subject dreams he is dead. It is worth noting that the patient suffers from intense death anxiety. We do not have any other elements in this dream—it is significant as much as it is concise. Dreaming of being dead means coming face-to-face with the impossibility of representing one's own death. That the subject dreams of being dead does not mean that he represents *his own* death, but rather that he dreams the very impossibility of representing it. In fact, the subject does not *see himself* dead in the dream but rather feels alive. "Realizing that I am dreaming my death, I feel as though I have survived, I feel alive and experience a sense of relief that, strangely, immediately fades into disappointment upon waking, which happens naturally, smoothly." In dreaming of his own death, he realizes that he is alive: an inverted *cogito* is at work here—*I am where I do not think*. The subject constantly thinks about his own death, that is, about something that cannot be represented; thus, he ends up dreaming about it, not to represent it but to fulfill the unconscious desire to survive, that is, to assume its unrepresentability as a condition of the life of the living. Relief appears in the dream, where the tangible limit of representation (*Vorstellung*), "I dream of being dead/therefore I know I am alive," frees the unconscious realization at the level of *Darstellung*: the subject is pushed out of his cogitating position, which is that of a detached observer—essentially dead—of his own life and instead touches the paradox of immanence to *one* life. The satisfaction achieved reveals that behind the anxiety of death lies an anxiety of life: upon waking, the relief vanishes because the subject, to put it in Lacan's terms, by waking up goes back to *dreaming with open eyes* in the realm of representation.

These two dreams show us the four elements that articulate the structure: the sovereign signifier, the empire of signs over which it reigns, the subjectivation that occurs at the edge of signification, and finally, the object that writes the trajectory of satisfaction in the realm of the senses. At the center, we find the Thing, *cloaca maxima*, which, through the unconscious's demonstrated ability to make use of the empire of signs and its master signifier, can undergo its own morphogenesis: it becomes an *objet*, arranging the inside and the outside that are dynamically and reciprocally generated, and thereby producing the other satisfaction that emerges as the reverse side of the deadly dissatisfaction of the signifier.

The empire of signs is founded on this exceptional signifier, which is rooted precisely in the absolute enjoyment it seems to promise. This is why Lacan rightly links *The Empire of Signs* by Roland Barthes to Ōshima's film *In the Realm of the Senses*: it is the dream of absolute enjoyment, of a power that is infinitely *beyond*, that dominates the imperial order. A dream imposed by signification upon the living and its enjoyment, culminating in its ultimate fulfillment in the act of castration, as Ōshima teaches us.

At the center of command, we find the phallus in its dual role of signifier of signification and illusion of absolute enjoyment. It is the One of sense/senses that the phallus makes present. And it is this One that dominates the commandment of transference in the form of the eternal dream of awakening, of liberation, as access to the scene of an absolute enjoyment of the law. Where the phallus structurally fails, there is discomfort, the symptom that psychoanalysis intercepts: the disjunction between knowledge and power, the realization that there is no guarantee, no absolute, no Other of the Other. And it is here that, in place of rogue nihilism, the ethics of the analyst imposes an additional step: beyond the One and its cynical failure, the real awaits the act that will allow the subject to find, in the object dearest to him/her, the ethos of another approach. Where the empire falls, subjectivation faces a decisive ultimatum: it must transform the nonexistence of the Other into an opportunity for a new *Weltanschauung*, characterized by the cynical trait that inaugurates it. This means treating this nonexistence as that which leads to the realization, as Lacan says, that "There is no neighbor if it is not the very hollow that is in you, your own emptiness" [19, p. 16]. It is necessary therefore to return to where we began: behind the numerical succession of signifiers, the decisive point is the passage from 0 to 1. What initiates the signifying machine is the madness of the instant of decision, as Kierkegaard emphasized. The ethical dimension of subjectivation resides there, in the traumatic point of grafting, where "confronted with the primary signifier, the subject is, for the first time, in a position to subject himself to it. There only may the signification of a limitless love emerge, because it is outside the limits of the law, where alone it may live" [17, p. 276].

It is precisely here that human beings have shaped the veil that separates them from the real: the narcissistic screen offered to them by the image of their individual body.

4.2.3 The Narcissistic Deviation

Let now return to the point of attachment, the event of the living being's inscription in the chain. Between 0 and 1 something always escapes us, and it is here that, horrified, not wanting to know, we cover it with the imaginary screen of the imago of one's own body. Lacan, in fact, entered the psychoanalytic debate precisely by emphasizing the importance of the imago of one's own body in the process of subjectivation and, at the same time, its function as a defense against the enigma of castration. We imagine the beautiful shape of the imago of one's own body so as not to confront the hole of the real body, which is inscribed precisely between 0 and 2, between the superegoic signifier of the empire (of signs) and its impact on the living being (in the senses). A double inscription or, to use Jean-Luc Nancy's term, a diplopia of *excription* (see [40]): the signifier impacts, carves, writes and erases, outlines, borders, tattoos the body, and in doing so, makes it exist as that which is stealthily extracted from it. The imaginary thus seals the wound of the signifying empire: *the tenderness of the hand that caresses the body pierced by the signifier articulates the affective chromaticism of the human being.*

At the point where the living being is captured in the signifying chain, a cut is produced, which the imaginary tends to saturate by offering us, through the image, a solution that compensates for the traumatic absence of the Other. It is a senseless wound, a cut—trauma—without cover: the imaginary acts as a seal and allows the deadly enjoyment of the signifier to encounter, instead of the despotic fate and its fatal outcome—the castration staged by Oshima—the possibility of a longer *detour*. Here too, defense could not take hold of the living being if it were not for the satisfaction it entails. It is not only about the image of the body that acts as a veil against the real of the hole, but also the fact that we love that image. Once again, we find transference, this time on an imaginary level. Though we must not forget that for Lacan, the root of transference, whether imaginary or symbolic in its articulation, is the real: "The reality of the unconscious is sexual reality—an untenable truth" [17, p. 150]. The narcissistic image of our body protects us, but at the same time it captures us in the transferral enchantment of its imaginary seizure because it offers us a certain degree of satisfaction. This is what Freud had already emphasized in the connection between narcissism and autoeroticism. The ego, therefore, is not only an object that serves as a screen against the enigma of castration (which ultimately represents the enigma of the Other's desire, as we will see), it is also an object of enjoyment.

At the point of capture, the human being is she/he who turns the wound inflicted on the living into an opportunity for satisfaction. This operation allows him/her to produce a share of *value* that begins to circulate within the body with the aid of the specular image. This is where autoeroticism comes into play, making the ego the privileged object of its own satisfaction. The image of the body thus performs two functions simultaneously: it projects the real body onto the aesthetic screen of the imaginary, disciplining its movement, mastering its motor capacity ("*maîtr*ise / *mot*or"), and regulating its behavior according to the ideal form befitting the "good man" [19, p. 264]; while at the same time it encloses within itself a certain share of enjoyment, which grants value to the imaginary by virtue of the autoerotic satisfaction it guarantees.

This is how the fundamental disjunction between knowledge and power is produced. The imaginary offers us, through the autoerotic mastery of the body and its ideal form, the sense of power and control that we exercise over ourselves, the world, and others. It is the power of the phantom that confers meaning where signs constantly expose it to the drift of meaninglessness inscribed in the senses and their point of collapse. On the other hand, knowledge becomes detached from power and roots itself in the point of capture, persistently questioning the fate of the Other's desire, whose trace has tattooed the body and whose traumatic wound continues to insist despite the imaginary remedies attempting to saturate it. The autoerotic solution that the power of narcissism offers leaves unanswered the epistemic question of the enigma of castration, of the Other's desire: *what do you want? Value* wavers, power yields to impotence, hysteria shakes the foundations of the "paper tiger" (Ibidem, p. 280) with which the imaginary veils the phallic enigma. Real dissatisfaction then takes over, suddenly revealing the human being's double failure: on the symbolic level, exposing him/her to the interdiction of the empire that denies access to enjoyment, which nonetheless exists symbolically elsewhere; on the imaginary level, exposing him/her to the *foreclosure* that conceals the enigma of castration and, at the same time, access to the Other's desire, leaving him/her stuck in the point of inhibition that the beautiful form of the narcissistic body generates when confronted with the real of the encounter. The point of attachment is a point of denied, erased satisfaction—it is a point of original repression. And it is on this point that we must insist.

4.2.4 *The Hu-Man Being (*L'être Hu-Main*)*

The ethics of psychoanalysis offers us the possibility of shifting away from this double failure, both imaginary and symbolic: to lose the power that the imaginary grants us through its narcissistic and autoerotic enjoyment; to renounce the aspirations of knowledge, which, absorbed by the power of the chain, forgets its substructural point of attachment. It is a matter of insisting, of lingering at the point of inscription, of the trauma and the event of the structural capture of the living being. It means learning to endure the (imaginary) impotence that this entails together with the not-knowing to which it exposes us. This is precisely what Lacan offers us by replacing the object that post-Freudians conceived to alleviate the suffering of Freudian castration with his original invention: "*objet a.*" This object does not provide us with a quick, imaginary escape from the empire of signs, yet neither does it not leave us entirely at the mercy of its commandment without ethical orientation. Indeed, it invites us to better understand the nature of the lost object that Freud handed down to us, because this, ultimately, is the real legacy of his work. As Lacan emphasizes: "If Freud draws our attention, it is not for what he thought as an individual at a certain turning point in his industrious life; what matters is the object he discovered" [18].

If we look at the object through a magnifying glass, we quickly realize that it is not something that once existed and was later lost, but rather an object that is structurally given as lost from the very beginning. The maternal breast, in this sense, is paradigmatic: it has never been encountered *as such* but always as the object of oral satisfaction, which from the very start is veiled by our imaginary narration that turns it into the sign of a mythical and primordial enjoyment—an absolute one, in fact. Yet, if we look more closely, the infant immediately experiences it in terms of "an eternally missing object," that is, a hollow:

> (…) this object, which is in fact simply the presence of a hollow, a void, which can be occupied, Freud tells us, by any object, and whose agency we know only in the form of the lost object, the *petit a*. The *objet petit a* is not the origin of the oral drive. It is not introduced as the original food, it is introduced from the fact that no food will ever satisfy the oral drive, except by circumventing the eternally lacking object [17, p. 180].

The *objet a* is a hollow, a contoured void. It is here, between the signifier and the living being, that the *quantum* of enjoyment (eternally lost) is written, which the object signals. It is here that the subject's singularity is ethically found—its original otherness, that peculiar estrangement that leads us to say, with Lacan, that the subject is *internally excluded* from its object. In other words, ultimately, it is precisely where the subject reveals itself as the thing most foreign to it and at the same time most intimate.

Here, finally, the signifying nature of the human being is manifested: of the subject that with its hand signifies its own place, the position it occupies in the world with its body, erasing its rooted localization to turn it into the possibility of delicately rewriting the edge, the contour, the boundary. The territory thus becomes the sedimentation of the object that we are: the continuous rewriting of the border of the eternally missing object, which insists in keeping the living being alive, caught in the chain of its signifiers. This is the meaning of Lacan's aphorism we began with: *the letter is the germ cell of life*. The writing of the border of that empty object, of the hollow, is literal—it sustains, through absence, the proper place of the "hu-man being" (*l'être hu-main*). The first signifier is, therefore, "the hand," with which one pragmatically and continuously draws the empty object, moving tentatively around its borders, skirting the archipelagos it ceaselessly arranges in the trajectory of the act with which we chart the imaginary cartography of our circumnavigation of the Thing.

The hand grasps and drops at the same time: to grasp, it must create emptiness, a hollow within itself. And so, the void welcomes the Thing that has become object. With the same gesture, the hand signifies the Thing and also embeds it in its corporealization. The signifier must therefore be understood as a trace, as the event of writing that continuously erases its own place, *expropriating it, altering it*. The human is the living being that has lost its place, because with the hand, it erases and incessantly incorporates that world which, as Lacan says, it constitutes "by committing suicide" [34, p. 124]. This is how it brings into existence the object nothing, or almost, which defines what is most essential to it.

Another satisfaction then flows through the hands, between the fingers, mine and yours, in the missed and therefore fulfilled encounter, in the touch, in the caress with which we keep tracing the untouchable that acquires consistency between us, on the reverse of the point of contact. Instead of the print the animal leaves as proof of its place in the world, the human being is "he who replaces his traces with his signature" [19, p. 272]. It is by virtue of this substitution that, instead of mere presence, the "hu-man being" (*l'être hu-main*) is exposed to the ecstatic experience of love. In erasing the print it leaves, it brings into existence the object nothing, or almost, which begins to feed the demand for love—the desire to know about the desire of the other—insisting in the chain of signifiers and exposing it to its reverse, one step at a time.

Another satisfaction reveals itself here, at the level of this inverted topology of the Other, which unveils the substructural writing of its economic functioning. While interdiction and foreclosure brought about dissatisfaction, the *objet a* opens up—in the hollow of the hand, in the *mis-take* of the signifying grasp—"beyond what is seen" (Ivi, p. 273). It punctures the narcissistic image of the body, it carries out the primordial substitution of the Thing with the object, and introduces in the alternation of 0 and 1 a certain density that points to another love, another satisfaction.

It is here, in this in-between, that the human being is given the opportunity for a vision. Instead of the image of one's own body, of the autoerotic narcissism that the mirror offers, that stands before me like a face-to-face encounter, a new image comes toward me, overtakes me, and strikes me from outside. Where language falls silent, in the quiet of words and their empire, another image reaches me. It is no longer the mother's face organizing the phantasmatic protection with which I shield myself from the real that concerns me: "Thus, from the one who received the key to the world in the slit of the prepubescent, the psychoanalyst must no longer expect a gaze, but finds himself becoming a voice" [29, p. 254]. Reduced to mere voice, pure enunciation, the analyst, having become the *refuse of humanity*, must therefore reposition himself starting from his/her gaze, which, "Even when lacking, (…) is clearly the object that presents each desire with its universal rule, by materializing its cause, in binding to it the subject's division 'between center and absence'" [32, pp. 462–463]. It is my own gaze, now reversed, that reaches me—not from the past, but from the future. An image yet to come, that opens up beyond the seen, piercing the aesthetic order founded on the primacy of the "ideic character" and its "visual centring" [17, p. 7].

The eternally missing object that the hand constructs around the point of contact between the oral cavity and the body of the Other is made of nothing, or almost nothing. It is the effect of the original erasure that the signifier produces and also replaces with its own signature, its own style, its own tactile mark. This trait repeats itself in every gesture, becoming the trace that, through the act, renews the satisfaction in pure loss that sustains human life. This *quantum* of enjoyment circulates within the discourse of the Other, within the signifying chain: step by step, it is the stumbling that repeats and renews itself in the form of *a*. In place of the Other, which does not exist, the human being is confronted with its *inform*. With the object nothing, or *almost*, that gives consistency to the Other starting from the void that flows through its hands. It is this *quantum* that gives value to the way the discourse of the Other takes hold of the living being, structuring the neurotic drama that sustains the phantasmatic vision that supports it. Another vision opens up here, just before the point where knowledge/power takes hold. Something is lost, erased, which means something also lingers in the background, beneath the surface. Between the hand that grasps and the object that is grasped, the thin film of this hollow speaks of the *ethics of touch* that is written alongside the point of contact with the world, with the other, with *oneself*. Here in this in-between what is untouchable is neither absolute enjoyment, prohibited by the sign, nor the real, both imagined and at the same time inhibited by phantasmatic foreclosure. Rather, it is the empty object that, falling from the hands of narcissistic autoerotism, relinquishes its vacuity to the Other, suddenly revealing that the secret of its consistency lies precisely where it is continually lost: in the point of contact that is a point of fall, of cessation, where the one is two and the two is one. At that point, finally, autoerotism surrenders the secret of its own construction to the Other, *which does not exist* but which, in the act of faith of this cessation, consists, in that instant that is enough to bring about the grace of touch.

This "blind vision" is the germ cell of life.

References

1. Agamben, G.: *Infanzia e storia. Distruzione dell'esperienza e origine della storia.* Einaudi: Torino (1978) [Engl. trans. by L. Heron. *Infancy and History. The Distruction of Experience.* Verso, New York, London 1993]
2. Agamben, G.: L'immagine immemoriale. In: La potenza del pensiero, Neri Pozza, Vicenza (2005)
3. Alt P.-A.: Sigmund Freud. Der Arzt der Moderne. Eine Biographie, C.H. Beck, München (2016)

4. Aristotele: (Met.).Metá tá physiká theta. In: Metaphisics Book Theta, Clarendon Press, Oxford [Eng. Trans. by Makin, S. 2006]
5. Auger, P.: L'Homme microscopique: essai de monadologie. Flammarion, Paris (1952)
6. Barthes, R.: Camera Lucida. Reflections on Photography (1980) [Translated by Richard Howard. Hill and Wang]
7. Bonazzi, M.: The unary trait of psychoanalysis. Identification, transmission, generation. In: Panebianco F., Serrelli, E. (eds) Understanding Cultural Traits. A Mutlidisciplinary Perspective on Cultural Diversity. Springer, Switzerland (2016)
8. Bonazzi, M.: "Vociferazioni o del godimento della voce. In: Serra, C., Bonazzi, M., Vizzardelli, S. (eds.) Voce. Un incontro tra filosofia e psicoanalisi. OT-Orbis Tertius, Mimesis, Milano (2018)
9. De Certeau, M.: Psychoanalysis and its History. In: Preziosi, D., Farago, C. (eds.) Grasping the World. The Idea of the Museum. Routledge (2004)
10. Ferenczi, S.: Thalassa: a theory of genitality. Routledge (2019)
11. Freud S.: Three essays on the theory of sexuality in the standard edition of the complete psychological works of Sigmund Freud, Volume VII (1901–1905): a case of hysteria, three essays on sexuality and other works. The Hogarth Press and the Institute of Psycho-Analysis, London 1953 (1905)
12. Guattari, F.: La chaosmose schizo. Chimères. Revue des schizoanalyses, N°13, automne 1991, pp. 82–90 (1991)
13. Lacan, J.: The object relation: the seminar of Jacques Lacan, Book IV. In: Miller, J.-A. (ed.) (1956–57) [Trans. by Price, A.R.. Polity, Cambridge 2021]
14. Lacan, J.: Desire and its Interpretation: The Seminar of Jacques Lacan, Book VI. In: Miller, J.-A. (ed.) (1958–59) [Trans. by Fink B. Polity, Cambridge 2019]
15. Lacan, J.: The ethics of psychoanalysis: the seminar of Jacques Lacan, Book VII. In: Miller, J.-A. (ed.) (1959–1960) [Trans. by Porter, D. Routledge, London 1992]
16. Lacan, J.: Transference: the seminar of Jacques Lacan, Book VIII. In: Miller, J.-A. (ed.) (1960–61) [Trans. by Fink, B. Polity, Cambridge 2015]
17. Lacan J.: The four fundamental concepts of psychoanalysis. In: Miller, J.-A. (ed.) Hogarth, London (1964) [Trans. by Sheridan, A. 1977]
18. Lacan, J.: The seminar of Jacques Lacan: Book XIV: the logic of fantasy (1966–67) [Unpublished translation by Cormac Gallagher. Session of May 10, 1967]. Retrieved on 25 Feb 2025. http://www.lacaninireland.com/web/wp-content/uploads/2010/06/14-Logic-of-Phantasy-Complete.pdf>
19. Lacan, J.: From an other to the other: the seminar of Jacques Lacan. Book XVI. In: Miller, J.-A. (ed.) (1968–69) [Trans. by Fink, B. Polity 2023]
20. Lacan, J.: The other side of psychoanalysis: the seminar of Jacques Lacan, Book XVII. In: Miller, J.-A (ed.) (1969–70) [Trans. by Grigg, R. Norton, New York, London 2007]
21. Lacan J.: Television: a challenge to the psychoanalytic establishment. In: Copjec, J. (ed.) (1974) [Translated by Hollier, D., Krauss, R., Michelson, A., Mehlman, J. Norton, New York 1990]
22. Lacan, J.: Lettres de L'École freudienne. La Troisième, no 16 (1975)
23. Lacan, J.: Le rêve d'Aristote. In Aristote aujourd'hui. ERÈS-UNESCO, Paris (1988)
24. Lacan, J.: Note to the Italian Group. Analysis **7** (1997)
25. Lacan, J.: Encore, on feminine sexuality, the limits of love and knowledge: the seminar of Jacques Lacan, Book XX (1972–1973). In: Miller, J.-A. (ed.) Norton, New York (1998) [Translated by Fink, B]
26. Lacan, J.: La méprise du sujet supposé savoir. In Silicet, 1, Éd, du Seuil, 1968, pp. 31–41; Republished In Autres écrits. Éditions du Seuil, Paris (2001a)
27. Lacan, J.: Joyce le symptôme. In: Autres écrits, Éditions du Seuil, Paris (2001b)
28. Lacan, J.: La psychanalyse. Raison d'un échec. Autre écrtis, Éditions du Seuil, Paris (2001c)
29. Lacan, J.: Proposition du 9 octobre 1967 sur le psychanalyste de l'École". In: Autres écrits, Éditions du Seuil, Paris (2001d)
30. Lacan, J.: The mirror stage as formative of the I function as revealed in psychoanalytic experience (2006a) [Trans. by Fink, B. In: Écrits: the first complete Edition in English, 75–81. Norton, New York]

31. Lacan, J.: Science and truth (2006b) [Trans. by Fink, B. In: Écrits: the first complete Edition in English, (pp. 726–45). Norton, New York]
32. Lacan, J.: Kant with Sade(2006c) [Trans. by Fink, B. In: Écrits: The First Complete Edition in English (pp. 645–68). Norton, New York]
33. Lacan, J.: The function and the field of speech and language in psychoanalysis (2006d) [Trans. by Fink, B. In: Écrits: The First Complete Edition in English (pp. 197–268). Norton, New York]
34. Lacan, J.: Aggressiveness in psychoanalysis (2006e) [Trans. by Fink, B. In: Écrits: The First Complete Edition in English (pp. 82–101). Norton, New York]
35. Lacan, J.: Le phénomène lacanien. In: Conférence prononcée au Centre universitaire méditerranéen de Nice le 30 novembre 1974. Texte établi par Jacques-Alain Miller, tiré à part des Cahiers cliniques de Nice, no 1, juin 1998, Section clinique de Nice (2011)
36. Leibniz, G.W.: Principia philosophiae. In: Monadology.Edinburgh University Press, Edinburgh (1714) [Translated by Strickland, L. 2014]
37. Miller J.-A.: C.S.T. Ornicar?, no 29 (1984)
38. Miller J.-A.: Biologie lacanienne et événement de corps. La Cause freudienne, no. 44 (2000a)
39. Miller J.-A.: L'orientation lacanienne 3,1. In: La Cause freudienne, no. 44 (2000b)
40. Nancy J.-L.: Corpus (1992) [Trans. by Rand, R.A. Fordham University Press, New York 2008]
41. Pfauwadel, A.: Lacan versus Foucault. Cerf, Paris (2022)
42. Sarti, A., Citti G., Piotrowski, D.: Differential Heterogenesis. Mutant Forms, Sensitive Bodies. Springer, Berlin (2022)
43. Stevens, A.: L'adolescence, symptôme de la puberté. Feuillets du Courtil 1998, pp. 79–92 (1998)
44. Stiegler, B.: Nietzsche et la biologie. PUF, Paris (2001)
45. Thom, R.: Modèles mathématiques de la morphogenèse. Recueil de textes sur la théorie des catastrophes et ses applications. Union générale d'éditions, Paris (1974) [Trans. Mathematical Models of Morphogenesis. Ellis Horwood Ltd. 1984]
46. Zupančič, A. What IS Sex? Massachusetts Institute of Technology (2017)

Chapter 5
INFANS

Marco Mazzeo

Abstract The essay offers a radical critique of the recapitulation theory, showing that human ontogeny is not a miniature of phylogeny but an autonomous and transformative process. Infancy is conceived as a structural condition of the human species, marked by neoteny: a premature birth and delayed maturation that expose the subject to prolonged evolutionary, cognitive, and historical plasticity. This dual nature—early fragility and extended potential—makes ontogeny a driver of history, rather than its outcome. The temporality of infancy breaks with evolutionary linearity and opens up to discontinuous time, capable of invention and rupture. The infant, without speech and yet bearer of ambivalent potentialities, becomes a figure of exposed, historical, and unfinished subjectivity, continuously rewritten through relations with others and with language.

5.1 Against Haeckel: Evolution as a Tv Series

> This paleontological history of the development of organisms, which we may term Phylogeny, stands in the most important and remarkable relation to the other branch of organic history of development, I mean that of individuals, or Ontogeny. On the whole, the one runs parallel to the other. In fact, the history of individual development, or Ontogeny, is a short and quick recapitulation of paleontological development, or Phylogeny, dependent on the laws of Inheritance and Adaptation.

With these words Ernst [18, pp. 10–11], describes the recapitulation theory in one of his best-known works. Since the 1920s [15, p. 127], this theory has started to be questioned. At the same time, however, still today it is still highly considered in the most diverse fields including psychology, anthropology and philosophy.

Probably, a reason of such success is that the recapitulation theory proposes a vision of gradual and progressive life on earth. Using a metaphor taken from the contemporary world, according to Haeckel, when observing the growth of an

M. Mazzeo (✉)
Università La Sapienza di Roma, Roma, Italy
e-mail: marco.mazzeo@uniroma1.it

S. Vizzardelli and R. Panattoni (eds.), *Plasticity and Discontinuity*, Lecture Notes in Morphogenesis, https://doi.org/10.1007/978-3-032-03119-8_5

organism, we could be eyewitnesses of the stages that the species has gone through so far, just like when, watching a TV series, we have the chance to watch the recap of the previous episodes in order to pick-up the thread of the story. Both cases can be applied to what for the German biologist constitutes a fundamental principle: "ontogeny itself is parallel only to a *portion* of phylogeny" (Haeckel [18, p. 314], p. 314. Italics in the text). Ontogeny repeats the *highlights* of the evolutionary history that predates it, and the highlights only, because the historical prelude for individual life would be, otherwise, too long and cumbersome.

The recapitulation law is secondly a progressive law. To an evolutionary episode another must follow that constitutes its development and explanation. In the same way a TV series progresses towards the narrative climax of the plot, so evolution should follow what, according to Haeckel (ibid., p. 282) is "the second great fundamental law" for biology but also for "the history of nations" (ibid.): "the progressive development" (*ibidem*). In what Haeckel calls "biogenetic law" (for a discussion in light of the most recent research: [20], an additional principle related to the biologist's adherence to Lamarck's theory of inheritance of acquired characters (proposed by Lamarck) is applied: changes acquired during the adult stage are recorded in the species as last episode of the evolutionary series.

We are referring to a Lamarckian theory that has been scientifically questioned for over a century, often being accused to have provided evolutionary material to the eugenic ideas that will triumph during the Third Reich (for a discussion see [31]). Why does the recapitulation theory constitute, still today, an implicit theoretical cliché?

I propose the following hypothesis: the apparently intuitive nature of this idea lies in an interpretation of both evolutionary and historical time, in terms of a progressive line (time) consisting of unextended geometric points that cannot be transformed, that is, adult individuals. Following the footsteps of Lamarck and Goethe, Haeckel laid the basis for what we could name *biological historicism*. Both Darwinian evolution and the human history would be characterized by their cumulative linearity. This is the reason why the interest on ontogeny is shallow: as we will see ontogeny itself is the evolutionary driving force capable of shattering this idea. Certainly, Haeckel, on the contrary (Ibid., p. 293) claims that "cannot indeed but be astonished when we consider the deep ignorance which still prevails, in the widest circles, about the facts of the individual development of man and organisms in general". However, truth is, that "his primary interest lay in the tracing of lineage" [15, p. 84]. As a matter of fact, according to the German biologist, "species transformation -given population pressure and struggle for existence, variability of traits, and inheritance of modifications- had to occur as a matter of mathematical necessity, so that really no further proof was required" [31, p. 228].

The recapitulation theory is the enemy to infancy because the infant would only be a distant echo of what once was the early adult human. Already in the prenatal stage "the embryo goes through the "adult" morphological stages of its ancestral progenitor" (Ibid., p. 154). In nonhuman species too, the cub would not only limit itself to represent the movie that has predated its morphogenesis, but it would be the biologic version of a film that portrays only adult predecessors. Unlike other versions of the recapitulation theory (for example that of Karl Von Baer: (Ibid., p. 149); [15, pp. 53–63, 109–114], according to Haeckel ontogeny shall be a prequel to the adult life that, in turn, only refers to adults.

At the end of the 19th Century, Haeckel addresses his theory to the idea of the superiority of the white race and the European nations (in the first place Germany). According to [31: p. 248 and ff.]: beautiful theoretical biography, the researcher was so sensitive to his contemporary history, that in the 8th edition of *The Natural History of Creation* he modified the position of the Japanese people in the stem-tree of the human species. Native Indian Americans went back in the standings because they had been defeated by the European settlers; they were sidelined by the Japanese people because "a new constitution for the Japanese nation was constructed explicitly on the German model and promulgated in 1889" (Ibid., pp. 249–250).

Today, such awareness related to the context of the recapitulation theory is quite different. In the neoliberalism universe inhabited by billions of Robinson Crusoe types (that is, human beings who declare themselves as individuals who are born independent from other individuals), Haeckel's theory goes through a second renaissance in the name of a species considered as an archipelago of already adult "island-human beings". The alleged impossibility of the ontogenetic transformation becomes implicit foundation of the alleged impossibility of the transformation of our history.

5.2 The Curious Case of Michael Tomasello: The Comparative Approach to the Recapitulation Theory

As compared to the 19th Century, the scenario has completely changed. Today, the progressive views of evolution are considered antiscientific, thanks to the synthesis between genetics and Darwinian evolutionism many of the mechanisms that only few decades ago seemed to be mysterious have been clarified (for a more detailed overview: [16]). Over the last decades, the science of Evo-Devo has been fostered "the comparison of developmental genes between species", "a new discipline at the interface of embryology and evolutionary biology" that, among other things, has revealed "a common "tool kit" of "master" genes that govern the formation and patterning of their bodies" [10, p. 10].

On the other hand, especially in the field of study of *Homo sapiens*, the current interest in ontogeny seems to be still limited. Confined in the specific field of pedagogy and education science, it has been struggling to make its way in a wider and more ambitious theory of human nature. According to Anthony Pellegrin [28, p. 38]: "Past 30 years have witnessed a renaissance of evolutionary approaches to the study of human behaviour and cognition, whether under the guise of sociobiology (e.g., Dawkins 1976) or evolutionary psychology (e.g., Buss 1989; Cosmides and Tooby 1987)", claiming also that (*ibidem*): "these approaches have paid little attention to the importance of ontogeny".

In order to check the persistence, if not of the recapitulation theory itself in its original version, of a biologic historicism similar to Haeckel's approach (evolutionary and historical time as a cumulative progress-oriented line), it is necessary to meet quite a complicated challenge. It is necessary to check its presence where you don't expect to find it. In this regard, the evolutionary theory of Michael Tomasello seems to represent a perfect case study. The theory has a particularly interesting approach because it has a critical approach towards the strictest dictates of the evolutionary psychology, according to which "new adaptations only arise in response to specific ecological challenges" with a "mostly conservative" idea of "evolution" [39]. Since the 90's Tomasello's team has been carrying out a huge amount of experimental and theoretical research work, open to the contribution of authors, who are not usually present in this context such as Ludwig Wittgenstein and Lev Vygotsky, particularly sensitive to a child-centred anthropology (the psychologist refers to his theory as "Neo-Vygotskian": [39, Chap. 9].

Premises are good. Tomasello dedicates an entire volume to ontogeny [39]; since his first monographies he has been insisting on the fact that human ontogeny is an "extended ontogeny" [37, p. 212], "with much individual learning" (*ibidem*) underlying two fundamental processes, that is, "sociogenesis" and "cultural history" [39, p. 47].

After this premise where "Cognitive scientists' underappreciation of ontogeny and its formative role in the creation of mature forms of human cognition" [37, p. 48] is criticized, Tomasello's analysis has ended up using some theoretical assumptions that seem not to have gotten rid of Ernst Haeckel's ghost. We could call them the principles of the so-called "comparative approach of the recapitulation theory":

(1) Ontogenies of the various organisms are different from each other, however cognitive tests of children can be compared to those of adult great apes. In *A Natural History in Human Morality* [38, p. 42] Tomasello claims: "Methodologically, although there are no beings alive today identical to the earliest moralists we are envisioning here, we may nevertheless cite as relevant various lines of empirical research with young children". Therefore, human infancy would still be the mirror of our evolutionary past.
(2) In addition to this, [38, p. 35] claims: "And so let us imagine the last common ancestor of great apes and humans on the model of contemporary chimpanzees and bonobos". The evolutionary past would follow a linear progression (rather than a tree-model or, according to more modern Darwin's theories a bushy model). In turn, great apes are compared to the last common ancestor to the line that separates humans (homo *sapiens)* from chimpanzees e bonobos about 6 million of years ago. The sum of points (1) and (2) makes the comparison between young children, great apes and our adult ancestors, possible. Human infants are not the miniature version of the primitive adult Homo Sapiens species anymore, but they the double of primitive human forms of life that are classified as species of *Homo* not yet *sapiens* (the most quoted one is *Homo Heidelbergensis* dating back to about 400,000 years ago: cf. eg. [38, p. 15], [40, p. 92]. The logic is similar to that of Haeckel. Tomasello (Ibid., p. 23) claims it in his most

> recent monograph: "Chimpanzees are as good an ape as any to represent early great apes, and for early humans, we can make some analogies to contemporary human children".

It is not sufficient to specify immediately after that we are aware of "arbitrariness of my choices for model species" (Ibid., p. 25). The problem doesn't stem from the fact that the model is arbitrary but from the fact that it's not clear how *bonobos and chimpanzees could act as a model for our species.* The evolutionary distance between current apes and common progenitors with the *sapiens* line is of six million years: thinking that this timespan hasn't affected these animals isn't arbitrary; on the contrary it is an anti-evolutionary assumption. The also partial analogy between hominids and infants is considered valid because, since ancient times it has been claimed that our species would be that of an *homo economicus* according to which assumption there are "individual self- interest in capitalist markets" [38, p. 158]: that is not competitive but cooperative, according to Tomasello's theory (*ibidem*). The problem is that, as is well known, the market economy is a very recent historical innovation that at most dates to some centuries ago.

5.3 Human Neoteny

In several circumstances, Tomasello's research seems to be close to a turning point (it is worth repeating it: it is perhaps the most advanced and bravest model currently present in the jagged world of cognitive psychology). For instance, according to different studies "Modern humans' very slow ontogenies are illustrated by the very slow pace of brain development as compared with that of other apes (and likely early humans; Gunz et al. [17])" and "chimpanzee brains reach 90 percent of adult size already by two years of age, whereas modern human brains do not reach that mark until eight years" [40, p. 106]. These data are already present in *Becoming Humans* [39, p. 26] underlying the specificity of our "extended immaturity" (Ibid., p. 24). A useful keyword for a philosophical change of pace is the adjective "extended".

What does it mean that our immaturity or, as it is said in other sources [38, p. 155], [40, p. 106], our ontogeny is "extended"? The reason of the extension can be dual. The first one is related to the addition of new evolutionary episodes to ontogeny. In a cumulative perspective of evolution, that is, in accordance with what we have called "biologic historicism", human brain and its abilities are conceived in a layered model, that is "as a kind of layered onion" [40, p. 9] in which a various amount of phylogenetic abilities are added. The model on the agency and the human morality that Tomasello proposes is explicitly of this kind. The idea has still recapitulation reminiscences: one of Haecker's axioms is abandoned, "a shortened and simplified repetition of previous patterns" [31, p. 155], nevertheless, ontogeny would be slower only because it would need more time to recapitulate the previous stages. The second approach overturns the logical connection: *human ontogeny isn't slow because it is extended, but on the contrary it is extended because it is slow.*

We have to work on a paradigm that insists on the "neotenic" trait [15]—i.e. that maintains a youthful appearance- of human ontogeny.

This is the paradigm that can finally break up with the recapitulation theory. Haecker acknowledges the ontogenetic slowness in humans however as Gould has remarked (Ibid., p. 82), the German biologist defines it as a "falsified history" because it would correspond to "concealment of the epitome of phylogeny" (*ibidem*). On the contrary, the cases that Haecker calls "heterochrony" are not the betrayal of phylogeny but its transformative driving force. The recapitulation theory following either Haeckel's theories or the comparative ones of Tomasello, postulates the substantial structural fixity of such ontogeny that would vary only in terms of evolutionary episodes that is capable to include. On the contrary, ontogeny is such precisely because it fixes the time of development and maturation of a single species.

Not only an organism has its own metabolic rate, partially related to body size [26], and a specific time capable of organizing its own perceptual environment [42]. Every species has its own body time: sexual maturation and biological aging.

In this regard human ontogeny is a "paradox" Barry [4, p. 200] because it is essentially focused on infancy. Gestation is accelerated: even though "in absolute terms, human newborns are roughly two times larger than great ape newborns both for total body size and for brain size" (p. 411), the length of human gestation is shorter than that of a human-size chimpanzee. According to Adolf Portmann [30, p. 57]: "the human is born almost a year 'too soon'" because "the actual length of human pregnancy is much less than it should be for typical mammalian development at our level of organization" (Ibid., p. 51). At the same time, after the birth and a first acceleration, rhythms tend to be delayed. In order to provide the readers with a general overview here are some more recent data: [15]: "the human brain reaches 95% of its adult size between 7 and 11 years of age and completes the final 5% of growth in a period of time nearly as long as that required for the initial 95%" [9, p. 730]; on average human sexual maturation of begins at the age 16.5 years versus 9.8 years of chimpanzees [8, p. 19]; the period of myelination of several areas of the neocortex, included the prefrontal cortex, is more extended compared to chimpanzees [27]. Furthermore, in human childhood, the volume of the white matter in this area is also larger [32]. MRNA expression levels of human prefrontal cortex show neotenic traits compared to those of chimpanzees and macaques [36, 21]. Always in the prefrontal cortex "the period of overproduction and elimination of dendritic spines on pyramidal neurons in this area extends to the third decade of life" showing "extraordinary neoteny" (Petaniek et al. [29, p. 13.281]. More generally, "The maximum period of synaptogenesis seems to be much longer in humans than in other mammals and the human brain undergoes a prolonged period of development before reaching the adult state" [9, p. 731].

What [15, Chap. 10] calls "neotenical matrix" opens human ontogeny to the category of potentiality twice (in response to the objections raised against Gould, see: [24]. As a matter of fact, such potentiality is characterised by a preterm birth making public, i.e. socio-historical, what for the other mammals is biologic, open to the relationship with other individuals what for primates is within the mother's womb and related to the biologic development. The subsequent ontogenetic delay in the retardation of sexual maturation, and more generally, of the processes of functional specialization extends predispositions toward learning that are generally limited during the first stage of the animal life. Preterm birth exposes individuals precociously to the need of learning to survive, the chronicity of infancy extends this need throughout the entire lifespan. Therefore, neoteny opens to a specific type of philosophical category of possibility that needs further research. According to a recent article it's not a matter of "individual pure (mere, non-actual) possibilities which are mind-independent as well as entirely independent of any actuality and actual reality" [14, p. 34]. On the contrary, it opens to the possibility that arises from a biologic time of a body in flesh and bones, that is the body of *Homo sapiens* that has to meet survival's issues.

Neoteny doesn't consist in building a set of possibilities preventively established by the species, that is, by an ecological niche. It doesn't consist in the pure possibility that would give "advantage and superiority" (Ibid., p. 31) to humans. This is why neoteny is not a very comforting and even less reassuring developmental phenomenon. "Chronic ability of learning"—this could be another temporary definition of neoteny- doesn't necessarily mean reduction of aggression that, thanks to greater abilities of inhibition and cooperation [7], would make us "a more docile orientation to deal with members of their own species, which was favoured by natural selection mainly for its enhanced social benefits" (Ibid., p. 271). "Learning" also means learning how to kill better, calmly, in cold blood or indirectly. "Inhibiting" also means inhibiting the identification with another human being who suffers or needs help. "Cooperating" also means cooperating in order to systematically destroy a group of people or entire ethnic groups.

5.4 The Relationship Between Infancy and History

Let's try to gain a greater terminological precision. Animal ontogeny is not the recapitulation of phylogeny: neither in Haeckel's Lamarckian terms, nor in Tomasello's comparative terms. Human ontogeny, and only this one, is infancy. In human ontogeny, in fact, the genetic *(-geny)* aspect definitely predominates over the ontological (*onto-*) aspect. At birth, the traits traditionally attributed to the *Homo sapiens*, language and erect posture are not present (let's just think about the classic definitions of human beings as "animals who possess reason" (*logos)* and "featherless bipeds"). Word and bipedalism are rather a possible achievement of development following the first year of life (the case of the so-called *enfants sauvages* shows the contingency of both).

Infancy is privation of potency and at the same time excess of potency. Privation, what Aristotle called "stéresis", occurs evidently in preterm birth: the newborn is unwary, defenceless and dependent from the others' care. The newborn neither can't talk nor can't walk. According to Aristotle: the baby's birth corresponds to "a feature which is natural but which something does not have" (Met. 1046a 32–33) and more specifically, "*when* it is natural" (ibid., 34, my emphasis). On the other hand, the chronicity of learning possibilities normally limited to the first stages of the animal life enhances the exceeding character of the possible. According to Aristotle, "metà logou" (that is, language-related) capacity (Ibid., 1046b 2) is a "capacity for opposites" (therein, 1046b 16) that can open to contradicting possibilities: "heat only for heating", that is *metà logou potency*, "while the medical craft for both disease and health" (Ibid., 1046b 6–7).

Infancy is privation (first aspect), infancy is ambivalence[1] (second aspect).

By virtue of the first aspect, human ontogeny cannot be restricted to the concept of ecological niche, that is, a type of environment[2] suitable for the species. Tomasello [39, p. 24] talks about "ontogenic niche" and "cultural niche" Tomasello [41, p. 106]; a recent theoretician of ontogeny uses the definition of "developmental niche" [7, p. 229] whereas others refer to it as "socio-cognitive niche" [43]. The neotenic and childlike features of human ontogeny make these definitions opaque if not even contradictory. In fact, according to Gilbert Simondon, French philosopher and radical theoretician of ontogeny [35, p. 328]: "The social thus appears as a reality that is quite different from the milieu with respect to the individual, we can speak of the social milieu only imprecisely and by expanding its meaning. The social could be a milieu if the individuated being were a simple result accomplished once and for all, i.e. if he did not continue to live by transforming". The lack of traits at birth identifying the species (language, bipedalism) comes together as a whole with the chronic presence of ambivalent capacities of individuation that is, the historical construction of a social individual. Ambivalent because they are *metà logou* capacities, therefore they are open to pairs of opposites: aggression and pacification, cooperation and individualism.

---

[1] Here we have to leave out an important aspect of the problem. According to Aristotle it's a weak ambivalence that respects the principle of non-contradiction (*Met.* 1051a 10–12), whereas for the anthropology of potentiality a strong notion of strong ambivalence capable of considering non-contradiction as a rule and not as a principle would be more adequate (for a defence of this speculation see: Mazzeo [22]; For the relationship between ambivalence and principle of individuation see: [35]).

[2] The recent paradigm of the "Construction Niche Theory" has proposed an ontogenetic version of the idea that "the niche construction is a very general process, exhibited by all living organisms" [13, p. 298]. All living organisms would modify the environment where they live. It's an interesting model that contributes to emancipating from a static and passive vision of natural evolution that is, only focused on gene transmission. At the same time, the radical insistency of the approach seems to create more a theoretical confusion rather than a real epistemic gain. The concept of "niche", in fact, is applied to any phenomenon: from the behaviour of ants to the historical changes of the human species ("cultural-historical niche": therein, p. 301). At least three problems emerge. Firstly, humans are not only builders but also destroyers with a potentially devastating impact both on the members of their own species and on other ecological niches. Secondly, the definition of "ecological niche" is not clear. If it generically corresponds to whatever surrounds any organism then it becomes a circular notion. What would *not* an ecological niche be? Thirdly, the definition of radical Darwinian continuism is necessarily contradicted by a set of following speculations. Our niche would be an "ontogenetic niche" (*ibidem*) similar to that observed among different species of primates, however "human niche construction is unusually potent" (therein, p. 298). Humans (homo *sapiens)* would be the "ultimate niche constructor" (therein, p. 308), so that "humanity's 'cultural niche' has allowed our species to settle in all corners of the world" (therein, p. 300).

What happens to the literally "no-speaking" infant (*infans*) when he/she starts speaking? Agamben has answered to this question [1, trans. 1993, p. 48] in an essay that represents a milestone of the 20th Century debate on human ontogeny: "the experience, the infancy at issue here, cannot merely be something which chronologically precedes language and which, at a certain point, ceases to exist in order to spill into speech".

In other terms: what is the relationship between the two sides of infancy, between privation at birth and ambivalent abundance of potentiality? The converter between impotency of privation and potentiality of ambivalence is called "history". Slowing down, that is, delaying, sexual maturation and critical windows of development, biological time is replaced by another type of time. In the range between the lack of the traits of a species at birth and the chronic presence of abilities, the historical time catches on. In contrast to the ideas of biological historicism, we are not talking about linear, progressive and necessarily cumulative time. Vice versa, human time is discontinuous. And it is discontinuous internally: it allows revolutions, resets and radical transformations (the end or the beginning of slavery, the creation of new calendars, the assimilation of a culture into another). Accumulation is only one of its possible ways of evolving. It is discontinuous externally too, that is, in relation to the time of Darwinian evolution. Any changes in history are fostered by the chronic capability of humans to learn, acquire new habits, create new life forms, without neither respecting the logic of the ecological niche, nor the logic of the mere natural selection.[3] Human beings can have an impact on the Earth similar to that of an entire geological era (let's think about the fortunate notion of "Anthropocene"); [12] had already noticed the human, and uniquely human, capability of starting processes of artificial selection.

We could claim that the two sides of human infancy are structurally reflected on the equally twofold and internally contrasting nature of our life. "Acute infancy" (0–1 year and a half year of age), based on a constitutively preterm birth, and "chronic infancy" (2–4 years of age) on a delayed maturation, determine the twofold nature of language, that is, according to Agamben's analysis [1, p. 51] of Emile Benveniste's antinomy, "between language and discourse". Speech corresponds to world of enunciation [1, p. 54] that "it permits each speaker to appropriate to himself an entire language" [5, p. 226] and produces new language uses. Language is the result of the "closed world of the sign" [1, p. 54], that is, the system of signs that makes Italian

[3] In order to avoid any misunderstanding, it is better to reiterate the concept. This doesn't mean that human neotenic infancy is a mysterious, non-biologic or unexplainable phenomenon. Just to give a couple of examples. The production of some transgenic laboratory *Rhesus* monkeys has acknowledged the importance of the human gene MCPH1 in order to increase the "ratios of immature cells" [34, p. 485] in a brain area which is responsible for higher cognitive functioning, that is "the frontal lobes" (therein, p. 484). A better visual recognition memory has been noticed too, capable of standing longer temporal delays (therein, p. 489 and ff.). Secondly, neural nets (called "ART2") have been recently developed, simulating learning neoteny-related mechanisms capable of responding to "so called Stability-Plasticity Dilemma": "ART2 network must be capable of plasticity in order to learn about significant new inputs and make the categorization structure less dependent on the input presentation order. It also must retain stable in response to suppress noise" (Chenet l. 2011, p. 408).

or Japanese languages, languages that are passed down from one generation to the other. Typical of acute infancy is the problem of starting to talk, with the risk of not being able to do it and being cut out from the speakers' group (from stuttering to the political obstacle of being able to speak in public). The Italian philosopher quotes Benveniste who writes: "the semiotic (the sign) must be RECOGNIZED; the semantic (discourse) must be UNDERSTOOD". The infant has difficulty being understood when he/she pronounces new enunciations, (mispronounced or polysemous words, holophrasis), conversely he/she has difficulty recognizing what is a sign and what distinguishes it from other things of the world, that is, "to perceive a correspondence what is there and what has been there before" (*ibidem*).[4]

Therefore, we have to be careful. The twofold aspect of infancy is not to be considered as a sequence of subsequent stages: at first the acute phase, which is subsequently replaced by the chronic phase. Preterm birth has a permanent—that is, not transitory—impact on human nature. Early exposure of the infant to the world, vital needs of other individuals, learning historical languages (no communication systems that are mostly genetically encoded) change permanently humans' life, even after the first or second year of life. Conversely, the chronic aspect of human infancy has its own autonomy because it is not based on the growth velocity at a foetal rate of the first year of life. On the contrary, it is based on the slowing down of the maturation processes. Acute infancy is not a mere prelude of chronic infancy; chronic infancy is not the mere continuation of the acute phase.

We find a similar twofold aspect in institutions and means of production. As a matter of fact, history has two possibilities. In every generation history can unfold through stratification passing down traditions, usages and customs (hunting, farming, market economy, Nation-state or anarchist communities) learnt over all the life stages (chronic infancy). However, in every generation, history risks to lose what has been in the past. Any type of language, if it is not learnt by the children of the following generation, becomes a dead language. Any political form, if it doesn't meet the interests of a new generation, that by definition starts from scratch (acute infancy),

[4] The essay is consistent with the idea that infancy "set up in language the split between language and *discourse* which exclusively and fundamentally characterizes the human language" [1] and with the idea that "the historicity of the human being has its basis in this difference and discontinuity" (therein, p. 52). At the same time, this essay differs from various theories that are present in Agamben's book. Schematically: (1) Agamben (therein, p. 55) conceives Saussure's language and Benveniste's "semiotic" as "the pure pre-babble language of the nature". Language would be per se comparable to animal communication systems (therein, p. 52) and only thanks to the discourse it would become historical and human. On the contrary, "dans la langue il n'y a que des différences" (Saussure, 1922, p. 166) is already historical and completely different from animal communication systems [5]. (2) This ambiguity has arisen from a controversial interpretation of Saussure (Mazzeo, in press) and from the fact that Agamben conceives infancy as a unitary and not twofold (acute and chronic) phenomenon as argued in this paper: the two sides of infancy correspond to the two sides of language (3) Finally, [1] focuses on the issue of the possible "abolition of history" on the part of human beings, therefore he detects in ritual and play a "binary system" (therein, p. 74) that creates and at the same time tends to destroy human history. However, a decisive anthropological issue is still pending: *the structural antagonism between infancy and history*, between the new generation of speakers and potential reset of institutions, customs and traditions that each time a social system has to face.

is doomed to end. Therefore, infancy is on one hand a condition of possibility of history. If humans were born talking and bipedal, there would be a unique language and therefore, essentially no language but a communication system that would be structurally similar to the bee waggle dance (according to Agamben: "such a man would thereby at once united with his nature; his nature would always pre-exist, and nowhere in it would he find any discontinuity": Ibid., p. 52).

On the other hand, infancy is the enemy of history because it represents what constantly threatens its existence. If we follow this twofold aspect, it is also possible to read the well-known within the human *praxis* proposed by Hannah [2, p. 177] On one hand, there is the capability of humans (*sapiens)* of "taking an initiative, to begin". In fact, "it is like a second byrth" (Ibid., p. 176) and it corresponds to preterm birth: "it is not the beginning of something but of someone" (Ibid., p. 177), "action as beginning corresponds to the fact of the birth, [...] it is the actualization of the human condition of natality" (Ibid., p. 178). On the other hand, human praxis has another side concerning "the achievement in which many join by "bearing" and "finishing" the enterprise, by seeing it through" (Ibid., p. 189). "Action" has a second fundamental meaning: keeping doing, extending the action indeterminately or nevertheless on a lomg-term condition. Infancy chronicity is *praxis* chronicity.

Focusing only on one aspect it means taking the risk of misinterpreting the structure of human nature and its plasticity. Tomasello's insistence on the presence of human cultural phenomena of a "ratchet effect" capable of accumulating discoveries and applications contrived by the previous generation represents a paradigmatic case of the problem. The intuition is obviously important, the ratchet effect determines a distinctive character of cultural phenomena in comparison to those that are genetically transmitted. However, if this character is considered in a unilateral way, there is a risk of distorting its meaning. *Praxis,* that is the account of tipically human actions (politics and war, economy and rebellion) is reduced to what Aristotle calls "kinesis", that is, motion. The Greek philosopher distinguishes two types of motion. The first type has a "limit" (*peras,* Met. IX, 1048b 18) because it is limited to a a specific goal, for example "making thin" (ibid., 1048b 19). When the goal is accomplished, i.e. you lose weight, the action stops. The second type of action is without rachet effect because it lasts indefinitely throughout the human life, for example "thinking" or "living well" (ibid., 1049a 23–25). Without a vision of infancy in neotenic terms, the meaning of "ratchet effect" [37, p. 5 ff.] is distorted: instead of indicating the possible, however not necessary cumulative character of historical phenomena, there is the risk of indicating the alleged intrinsic end of cultural forms as if they were equivalent to the building of a technical object [2, p. 140 ff.]. Once the house is built, the action stops because it has achieved its goal. However, what would the goal be to be achieved by a language, a Nation-state or a barter system? What would the achieved goal be, what type of language, State or barter system would stop in order to vanish afterwords?

References

1. Agamben, G.: Infanzia e storia. Distruzione dell'esperienza e origine della storia. Einaudi, Torino (1978) [engl. trans. by L. Heron. Infancy and History. The Distruction of Experience. Verso: New York-London 1993]
2. Arendt, H.: The Human Condition. University of Chicago Press, Chicago and London (1958)
3. Aristotele (Met.). Metá tá physiká theta [engl. trans. by S. Makin. Metaphisics Book Theta. Clarendon Press, Oxford 2006]
4. Barry, B: The evolution of human childhood and fertility. In: Hawkes, K., e Paine, R.R. (eds.) The Evolution of Human Life History. School of American Research: Santa Fe-Oxford 2006, pp. 197–230 (2006)
5. Benveniste, E.: Problémes de linguistique générale. Gallimard, Paris (1966) [engl. trans. by M.E. Meek. Problems in General Linguistics. University of Miami Press: Miami 1971]
6. Bertollini A.: (in press). Agamben Reader of Saussure. II. The Benveniste Case. ahiers Ferdinand de Saussure
7. Bjorklund, D.F.: How Children Invented Humanity. The Role of Development in Human Evolution. Oxford, Oxford University Press (2021)
8. Bjorklund David, F.: Human evolution and the neotenous infant. In: SybilL H., Bjorklund, D. F. (eds.) Evolutionary Perspectives on Infancy, pp. 19–38. Springer, Cham (2022)
9. Bufill, E., Agusti, J., Blesa, R.: Human neoteny revisited: the case of synaptic plasticity. Am. J. Human Biol. **23**, 729–739 (2011)
10. Carroll, S.: Endless Forms Most Beautiful: The New Science of Evo Devo and the Making of the Animal Kingdom. Norton, New York (2005)
11. Chen, Z., Wang, M., He, L.: Neoteny learning rule of ART2 network and its function. In: Sixth International Conference on Natural Computation, Yantai (China), pp. 407–411 (2010)
12. Darwin, C.H.: The Variation of Animals and Plants under Domestication. John Murray, London (1868)
13. Flynn, E.G., Laland, K.N., Kendal, R.L., Kendal, J.R.: Developmental niche construction. Dev. Sci. **16**(2), 296–313 (2013)
14. Gilead, A.: Neoteny and the playground of pure possibilities. Int. J. Hum.Ities Soc. Sci. **5**(2), 30–39 (2015)
15. Gould, S.: Ontogeny and Philogeny. Harvard University Press, Harvard (1977)
16. Gould, S.: The Structure of Evolutionary Theory. Harvard University Press, Harvard (2002)
17. Gunz, P., Neubauer, S., Falk, D., Tafforeau, P., Le Cabec A., Smith, T.M., Kimbel, W.H., Spoor, F., Alemseged, Z.: Australopithecus afarensis endocasts suggest ape-like brain organization and prolonged brain growth. Sci. Adv. **6**, 14eaaz4729 (2020)
18. Haeckel, E.: Natürliche Schöpfungsgeschichte. In: Reimer G., Berlin (engl. transl. by R. Lankester. The History of Creation. Henry King & Co.: London 1876) (1868)
19. Hrvoj-Mihic, B., Bienvenu, Th., Stefanacci, L., Muotri, A.R., Semendeferi, K.: Evolution, development, and plasticity of the human brain: from molecules to bones. Front. Hum. Neurosci. **7**(707), 1–18 (2013)
20. Levit, G.S., Hoßfeld, U., Naumann, B., Lukas, P., Olsson, L.: The biogenetic law and the Gastraea theory: from Ernst Haeckel's discoveries to contemporary views. J. Exp. Zool. B (Mol. Dev. Evol.) 338, 13–27 (2020)
21. Liu, X., Somel, M., Tang, L., Yan, Z., Jiang, X., Guo, S., Yuan, Y., He, L., Oleksiak, A., Zhang, Y., Li, N., Hu, Y., Chen, W., Qiu, Z., Pääbo, S., Khaitovich, P.: Extension of cortical synaptic development distinguishes humans from chimpanzees and macaques. Genome Res. **22**, 611–622 (2012)
22. Mazzeo, M.: Contraddizione e malinconia. Saggio sull'ambivalenza. Macerata, Quodlibet (2009)
23. Mazzeo, M.: When less is more: neoteny and language. Cahiers Ferdinand de Saussure **67**, 115–130 (2014)

24. Mazzeo, M.: Capitalismo linguistico e natura umana. Per una storia naturale. DeriveApprodi, Roma (2019) [Spanish trans. Capitalismo lingüístico y naturaleza humana. Por una historia natural. Tercero incluido: Barcelona 2022]
25. Mazzeo M. (in press). The Vanished Toy: Agamben reads Saussure. Cahiers Ferdinand de Saussure
26. McMahon Th. A., Bonner J.T.: On size and life. W.H. Freeman & Co.: New York (1983)
27. Miller, D. J., Duka, T., Stimpson, C. D.: Prolonged myelination in human neocortical evolution. PNAS **109**(41), 16480–16485 (2012)
28. Pellegrini A.D.: The role of development on evolutionary psychology: Tinbergen revisited. Psychol. Inq. **19**, 38–40 (2008)
29. Petaniek, Z., Judaš, M., Šimic, G., et al.: Extraordinary neoteny of synaptic spines in the human prefrontal cortex. PNAS **108**(32), 13281–13286 (2011)
30. Portmann, A.: A Zoologist Looks at Humandkind. Columbia University Press, New York (1990)
31. Richards, R.J.: The Tragic Sense of Life: Ernst Haeckel and the Struggle Over Evolutionary Thought. The University of Chicago Press, Chicago (2008)
32. Sakai, T., Mikami, A., Tomonaga, M., Matsui, M., Suzuki, J., Hamada, Y., Tanaka, M., Miyabe-Nishiwaki, T., Makishima, H., Nakatsukasa, M., Matsuzawa, T.: Differential prefrontal white matter development in chimpanzees and humans. Curr. Biol. **21**, 1397–1402 (2011)
33. de Saussure, F.: Cours de linguistique générale. Payot, Paris (1972)
34. Shi, L., Luo, X., Jiang, J., et al.: Transgenic rhesus monkeys carrying the human MCPH1gene copies show human-like neoteny of brain development. Natl. Sci. Rev. **6**, 480–493 (2019)
35. Simondon, G.: L'individuation à la lumière des notions de forme et d'information. Éditions Jerôme Millon, Grenoble 2005 (1958) [Engl. trans. by T. Adkins. Individuation in Light of Notions of Form and Information. Minnesota University Press: Minneapolis-London 2020]
36. Somel, M., Franz, H., Yan, Z., et al.: Transcriptional neoteny in the human brain. PNAS **106**(14), 5743–5748 (2009)
37. Tomasello, M.: The Cultural Origins of Human Cognition. Harvard University Press, Harvard (1999)
38. Tomasello, M.: A Natural History of Human Morality. Harvard University Press, Harvard (2016)
39. Tomasello, M.: Becoming Human. A Theory of Ontogeny. Harvard, Harvard University Press (2019)
40. Tomasello, M.: The Evolution of Agency: Behavioral Organization from Lizards to Humans. MIT Press: Cambridge (Mass.) (2022)
41. Tomasello, M.: Origins of Human Communication. MIT Press, Cambridge (Mass) (2008)
42. Uexküll J. von.: Streifzüge durch die Umwelten von Tieren und Menschen. Ein Bilderbuch unsichtbarer Welten. Fischer: Frankfurt a. Main 1970 (1933) [Engl. transl. by J.D. O'Neil. A Foray into the Worlds of Animals and Humans. Minnesota University Press: Minneapolis-London 2010]
43. Whiten, A., Erdal, D.: The human socio-cognitive niche and its evolutionary origins. Philos. Trans. R. Soc. B **367**, 2119–2129 (2014)

Chapter 6
INTENSIO

Alessandra Campo

Abstract This essay explores the concept of intensity as a key to rethinking life and thought beyond the logic of identity and contradiction. Through a genealogy that includes Leibniz, Bergson, Ruyer, and Deleuze, the intensive is presented as the cosmogenetic principle of the real: a force that precedes all representation and measurement, but is not inarticulate. Reality is process, but process proceeds through infinitesimal qualitative variations before quantitative ones. Feeling them is the task of a transcendental empiricism and a psychic mathematics. Differential calculus is the writing of an ontology of becoming rather than a tool of mathematical analysis, because intensity is the sign of the living. To admit this is to resolve the knot of dynamic structuralism by conjugating form and event, eternity and time.

6.1 Foreword

Differential heterogenesis, in the words with which its inventors present it in more than one work, is, or would like to be the avant-garde—the juxtaposition with Dadaism often occurs—of the new way of thinking discovered, in a happy simultaneity at a distance, by Newton and Leibniz on the basis of the results obtained, before them, by Galileo, Cavalieri, Barrow and others. The infinitesimal calculus is a new way of thinking because, thanks to it, mathematics has, for the first time, penetrated the immediate datum of both internal and external consciousness, namely movement, trying to formalise it. To achieve this, the use of special magnitudes was, understandably, inevitable: what Whitehead calls the "passage of nature" [71, pp. 54–55] is not the mere accumulation or juxtaposition of elements that can be quantified with extensive magnitudes. Since passage is neither composable nor decomposable with the aid of *extra partes*, it is quantifiable, rather, by means of other magnitudes, thanks to the other of magnitude, provided we devise new notations—the *partes non quantae*, the infinitesimals, the intensive magnitudes—and expand the thresholds of perception to the point of a calculation: the *mathesis intensorum*. This "second

A. Campo (✉)
Università di Urbino, Urbino, Italy
e-mail: alessandra.campo@uniurb.it

S. Vizzardelli and R. Panattoni (eds.), *Plasticity and Discontinuity*, Lecture Notes in Morphogenesis, https://doi.org/10.1007/978-3-032-03119-8_6

application of mathematics to nature", as Kant would define it when expounding the principle of the *Anticipations of Perception* in the *Critique of Pure Reason*, is indeed extraordinary. Not only does it strive to measure the non-measurable, contributing to the victory of a challenge inaugurated by irrational numbers, the quadrature of circle and the problem, long debated by medieval scholasticism[1] but originally formulated in Aristotle's *Categories* (8.10 b26) of the *intensio et remissio formarum*. Infinitesimal mathematics is, moreover, a "psychic mathematics", according to Bergson's lectures on time at the Collège de France in 1903–1904 [11, p. 269], or an "aesthetic mathematics", if one takes seriously the Leibnizian matrix of the discipline founded by Baumgarten and the fact that "aesthesis", already for Plotinus in the *Enneads*, means "consciousness". As such, it aspires to follow the generation of magnitudes by computing the unfolding of the "moment" with a writing that would give the continuous the precision of the concept. And this generation, Newton explains in *De quadratura curvarum*, also takes place in nature: one observes it every day in the motions of bodies even before one can trace it on paper, or screen, with the hand guided by thought.

The objects of infinitesimal calculus are flowing quantities, both nascent and evanescent. Its starting point is the intuition of an interior of movement: the *intension*, which can also be an *intentio*, of the *generatio*. This interior, variously referred to by the terms "impetus", "motione" or "momentum", is the cause of the appearance of something—a magnitude, a form or a being (if one ontologises infinitesimals and their products: finite extents)—from "nothingness": the *nihil privativum* of which Kant, in the *Anticipations of Perception*, offers the example of darkness and cold. Descartes had judged as confusing, and above all foreign to true science, namely geometry, the idea of a continuous progress from nothing to something,[2] but infinitesimal analysis did not merely bring this idea into clear focus by founding, for it, a new science. By betting on the fact that *est aliquid praeter extensionem, imo extensione prius*, the new mathematics implanted the presupposition of an inextension that generates all extensions [21, p. 70] in the pure material datum of space and time and so an absolute position with which the reality of the objects of scientific experience is positively founded [21, p. 70] in the heart of geometry. In other words, what, in *The Introduction to Metaphysics*, Bergson welcomes as the the most powerful of the methods of investigation at the disposal of the human mind [7, p. 70], the differential calculus, bets on the fact that there is something more fundamental than mass and position in bodies, something, writes Leibniz in *Specimen Dynamicum*, that even precedes them:

---

[1] FOR a classic history of this debate see [47]. For a brief history of the problem focusing on the so-called "Oxford Calculators", see [67]. We cannot but mention the contribution of Nicholas of Oresme, whose graphic representation of qualitative variations had important implications for the development of mathematical concepts and forms of geometric representation in the late Middle Ages and early Modern Age. On this subject, see: [48, 20].

[2] In the letter sent to Mersenne on 11 October 1638, Descartes says he doesn't believe what Galileo says, namely that falling bodies pass through all the different speed levels. On this point see [43] and [15].

> the force of nature implanted in all things by the Creator. This force [...] is endowed with a conatus or effort (*nisus*), such that it will attain its full effect unless it is impeded by some contrary striving. This effort often makes itself felt by the senses, but in my view reason shows that it is everywhere in matter, even when it is not apparent to the senses [...] it constitutes the inmost nature of bodies. For to act is the mark of a substance, and extension alone, far from itself constituting substance, is no more than the continuation or diffusion of a given effort-exerting and counter-straining (that is, resisting) substance. I recognize that all corporeal action arises from motion, and that motion itself comes only from other motion, either already in the body or impressed from outside [45, pp. 154–155].

For Leibniz, there is nothing real in motion other than its *momentum*, which must be founded on the force tending to change: that *inextensum* which is the *intensio*. Unlike the limits postulated by Eudoxus' and Archimedes' method of exhaustion or its modern version, the method of indivisibles, the differential calculus assumes infinitesimals as the productive principles both of the object and, to a certain extent, of the *aisthesis* (sensation-consciousness) one has of it. This means, as Maimon also recognised[3], that the progress from a nothingness or a zero, which is always relative, to any given reality coincides with the genesis of both the observed and the observer because the infinitesimal is not the negative limit reached by division, but a positive reality from which to start, on pain of resting victim to a merely trivial conception of the passage: the transport of a unity from one point to another in space. Apart from the infinite, the finite is blind, contracted on the illusory guarantee of "the one plus one", because iteration, in Chatelet's words, clings naively

> to a manipulable unity and thus pretends to become intoxicated with concreteness, forgetting to fall headlong into the demands of the next, which leaves it no respite: it never ceases to yearn for the next +1. It will see 1, 2, 3... *n* poplars, but never a *row of* poplars, because a row, deprived of the horizon that unfolded the similarity, immediately dislocates into 1, 2, n poplars, and *so on*: the poplars are at once too different and too identical, and once one has begun such a count, one is soon surprised to hope for a "true difference", always pursued *beyond* the finite [18, pp. 88–89].

By renouncing to take advantage of the development of things [l'enveloppement des choses] [18, p. 89] and depriving itself of a contact with its source—what Chatelet calls "horizon"—iteration becomes sclerotized. Of course, one could go and verify *on the spot* that the poplars are not really at the same distance. Yet, without the horizon, without that *mente concipere* that has made the fortune of modern science even according to Heidegger [35, pp. 62 ff.], no device alludes to it. This is why, with a superior effort of intuition not unlike what Einstein would later do by teleporting himself into the photon, Galileo, Newton and Leibniz placed themselves, albeit with some differences, in the body ready to fall, where "ready" means "while". Here, in the levity of gravitating to the ground while hovering in the air, the earth and the air are generated, *sub specie durationis*, by the movement that intends them, without it being possible to consider any space prior to this event as anything other than an undue abstraction, a circumstance that differential heterogenesis enhances and of which, indeed, it makes use as a precious methodological principle.

---

[3] For Maimon, too, the authentic genesis of the object and the subject, of the thing and its perception, is announced in the continuity of intensive magnitudes [49, pp. 290 ff.].

The infinitesimal is generative insofar as it is the instance, transcendental according to Cohen and Deleuze, diagrammatic according to Chatelet, capable of conferring reality, for thought, on the onset of the sensible datum and reality, for the senses, on the advent of an idea. In short: it is the first intuitive support available there where nothing happens as nothing is seen, both with the *natural* and the *supranatural lumen* [54, p. 54]. In *Difference and Repetition*, Deleuze makes it, for this reason, the principle of a quadruple genesis: that of the *extensio* in the form of schema, that of extensity in the form of extensive magnitude, that of *qualitas* in the form of matter occupying extensity, and that of the *quale* in the form of the designation of an object. Not the given, then, but that by which the given is given [24, p. 222], since everything that is given is given twice.

> The first givenness, the *Empfindung,* is mysterious. It is the type of given ness and given in which the mode of givenness is itself the given. Affectivity is both the impression's mode of givenness and its impressional content. It is the transcendental in a radical and autonomous sense. And then, this first given, which is always already given and presupposed, is given a second time in and through intentionality, as a transcendent and irreal thing, as its "vis-à-vis" [37, p. 17].

For Henry, Husserl's critic, the thing thought in the presupposition of intentionality must first and foremost be something else: something real and immanent; something purely subjective. Everything given to us is constituted, in short, represented. But the constituted is, first of all, necessarily nonconstituted. Everything is transcendent but the transcendent itself is, first of all, nontranscendent [37, p. 17]. The problem, however, is to know whether the arch-donation is actually realised through itself, i.e. autonomously, like the snake which, as it moves, carries its body and the wave it thus draws at the same time, or whether, in order to grasp it, it is necessary to resort, once again, to the only mode of manifestation to which Western thought has given its titles: the ecstatic dimensionality of phenomenality for whose opening intentionality is only a name among many. In fact, infinitesimal logic is dominated by the concept of an origin that precedes all reality and all matter [54, p. 54], extensionality never being anything other than a result that supposes, as such, a genesis. But, in attempting to presuppose a foundation for the phenomenon that is spreading and to identify the *point in time* which, when added to itself, will end up producing the object as a given ([53], p.196, our. transl. and italics), one would be mistaken to confuse the transcription of the first hint of movement detected by the senses, even numerical and algorithmic, with its actual incipit.

Bodies do not begin to fall when we see them fall, nor are ideas born and die when they appear and disappear from the field of our consciousness: unless we reduce the *esse* to the *percepi* even more radically than Bishop Berkeley did by dismissing infinitesimals as "ghosts of departed quantities" in his sarcastic *The Analyst* [12: § 16,§ 34], it must be admitted that everything that rises, mutates and perishes does so continuously, "always" rather than "suddenly", even if this, for reasons we shall have to see, does not disqualify the absoluteness—synonymous with unpredictability, aleatoriness and discontinuity—with which such events present themselves to us and which is not so different from that with which the unconscious seems to manifest itself, with a *Witz* or an *Einfall*, to consciousness. As much as noon stands at the zenith by way of the aurora, it is in fact from midnight that it arrives, and that it arrives, Baumgarten also acknowledges, without jumps [die Natur keinen Sprung macht aus der Dunkelheit in die Deutlichkeit] [2, p.15]. If one can speak of a leap, it is only with reference to our attention because the continuity between consciousness and its other, both external and internal, is not demonstrable in any way within the boundaries of this consciousness. In their readings of Kant's *Anticipations of Perception*, Philonenko has had the merit of insisting on this very point by reflecting on the experience of the falling body:

> After a while, the fall is translated into space and that is when we *see the* body falling. We can determine the magnitude of the fall by referring to the portion of space travelled. But we all know that the movement of the body does not begin at the very moment when we *see it*, when we have the *sensation* of it. Between the moment when the body appears to us to be at rest (=0) and the moment when it already appears to be travelling through space at a speed that can be determined (=1), a whole series of events has taken place that are beyond our perception: falling bodies pass through all degrees of speed. And these degrees constitute *the origin of the phenomenon captured in the sensation* [54, pp. 196–197].

The conclusion is not long in coming: nothing is more difficult to conceive than this origin of the sensible [54, p. 57]. A myriad of degrees or small movements silently compose the great and thunderous cascade of nature, cunningly eluding our perception. And it is because of this, by reason, that is, of the ineradicable presence of a threshold—the *Ebenmerklichkeit* fixed by the Weber-Fechner law as the *Urmeter der Psychologie*—that a certain genesis appears differential or a certain generation spontaneous. This does not mean that they are not, but that they are otherwise from the way we have hitherto thought of them. The purpose of this essay is to illustrate the nuances of this "otherwise" by attempting to clarify in what sense the first thing seen, heard, or thought of *is and at the same time is not* the first thing. Saying otherwise, we will ask what the gap between being and feeling-thinking is a sign of if the genesis of the object is, at the same time, genesis of the consciousness that one has of it. Is there another consciousness, or another subject, external or in any case independent, of this generation, with respect to which the gap makes sense? What is the value to be attributed to the transcendence of phenomenological consciousness and to the belatedness of knowledge that the gap between being and appearing, noumenon and phenomenon fatally emphasises, if *intensio*, conceived, as Leibniz invites us to do, *sub specie roboris*, is present everywhere in nature even if we do not detect it and therefore, although delayed in terms of knowledge, we are in some way always involved, contemporary to that of which, nevertheless, we only become aware later? After all, even the delay, distance and abstraction of which science is so often accused, with the bias of immediate experience, occurs in the same nature from which calculation, and the perception that is reduced to it in the new mathematics, would take us away, namely within the thing from which the symbol snatches us away. So, given that we are included even when we exclude ourselves to observe, what is the ultimate significance of the gulf between Henry's two donations and of that act which, sometimes, manages to possess—as Bergson says in the *Introduction to Metaphysics*—the absolute [7, p. 3], that is, intuition? And this absolute, besides, what is it? What relationship does it have with the absoluteness of the beginning and end of phenomena, with the suddenness of the increase and decrease of intensity?

These questions can only be answered by a mathematical psychology such as Leibnizian monadology, whose paths, it is clear, all lead to the infinitesimal. But perhaps even better answered by the "revised" monadology of one of his disciples according to Deleuze: Raymond Ruyer. The notions of "absolute survey", "primary consciousness" and "secondary consciousness" put forward by this monadology correct[4] in the light of the results achieved with the transition from a first-rate empiricism—that of Locke, Hume and Berkeley—to a higher one—that of James, Bergson and Whitehead, as well as from those drawn from the most promising sciences of the 20th century with which Ruyer never ceases to engage—*Gestaltpsychologie*, Quantum Mechanics, cybernetics, genetics and molecular biology—are so important for the understanding of morphogenetic processes and brain plasticity that their absence from the debates on these topics is puzzling.

Ruyer's psychobiology—a metaphysics of absolute forms where "form" means "formation" and "formation" means "consciousness"—is perhaps the privileged philosophical framework to allow the spearhead of the new mathematics to develop as a knowledge of *the inside from the inside* rather than, as the *mathesis intensorum* has been so far, remaining a knowledge of *the inside from the outside* (Galileo, Newton) or of *the outside from the inside* (Leibniz, Kant). However, before concluding in this direction, in what follows we will first deep Bergson's intuition of what makes the "mathesis of the moderns" new, namely the assumption of a soul of the movement, and, after emphasising the affinity between this assumption and the arguments put forward by Leibniz to defend the existence of small perceptions, we will demonstrate, in the light of the richness of the immediate datum of consciousness—Bergson's "duration", James' "pure experience" and Chatelet's "horizon"—the inadequacy of "intellectualistic" knowledge trying to refer to it.

6.2 The New Way of Thinking: The Infinitesimal Analysis

6.2.1 The Intense Interior of Movement

Differential calculus is a psychic mathematics because it assumes an interior, and thus a psyche, to movement [11, pp. 271–272]. This must be taken literally: the movement is no longer observed from the outside [10, p. 224], through space, nor is it considered for how it appears to the consciousness of the observer, that is, as a

[4] Or "correct". As early as 1938 Ruyer appeals to «what is eternally true in monadism [ce qu'il y a d'éternellement vrai dans le monadisme] [57, p. 127]. In the years that followed, however, he embarked on a genuine reformulation of monadology (Ruyer's monads are forms and not substances, they are extended rather than inextended, and they are not individuated like Leibnizian monads) until, in his last works, he explicitly emphasised the monadological heritage of his thought. On Ruyer's monadology see [22].

set of positions, or stops, between which a moving point passes tracing a trajectory. The new mathematics tries to enter the movement and to tune with the agent of which the movement is, at the same time, essence and manifestation, to take up the title of a well-known Henry's work [36]. Like Leibniz, Bergson also describes this "interior" as something that, in the present, preforms the future [11, p. 273]. And it is the same definition that Bergson gives of intensity in his 1889's *Essay*: a present contraction and a future dilation, the image of a virtual extension and, if one may say so, of a compressed space [3, p. 4]. Interiority, like intensity is in fact the same duration conceived as the interpenetration *se faisante* between different states each of which announces what follows and contains what precedes it in such a way that it is impossible to say exactly where one ends and the other begins. Each element, in this qualitative or continuous multiplicity, prolongs and blurs so admirably into the other offering the model of the mutual confusion and dependence between observer and observed that, even today, makes theoretical physics despair.

Indeed, the great discordance between intellectualistic logic and sensible experience manifests itself above all when the experience is that of the influence of one event on another: what James, sometimes, denotes with the German *durcheinander* [39, p. 31; p.158]. From a critical point of view, Kant himself confesses in the *Anthropology*, we cannot conceive how dissimilar things could have sprung from one and the same source [42, p. 70], because intellectualism

> Denies that finite things can act on one another, for all things, once translated into concepts, remain shut up to themselves. To act on anything mean to get into it somehow; but that would mean to get out of oneself and he one's other, which is self-contradictory, etc. Meanwhile each of us actually *is* his own other to that ex tent [38, p. 106].

Unlike in space—the atmosphere in which the intellect bathes [6, p. 223]—in duration there is no element that separates itself from every other element as concepts, or words, separate themselves from each other. Nor is there any part so small that it is not, at the same time, also a passage, a conflux; that it does not approach, also, the neighbouring parts while preserving the memory of the distant ones. This means, and this is James' conclusion in the pamphlet he dedicates to Bergson's anti-intellectualism, that

> No part there is not really *next* its neighbours; which means that there is literally nothing between; which means again that no part goes exactly so far and no farther; that no part absolutely excludes another, but that they compenetrate and are cohesive; that if you tear out one, its roots bring out more with them: that whatever is real is telescoped and diffused into other reals; that, in short, every minutest thing is already its hegelian "own other" in the fullest sense of the term [38, pp. 110–111].

In Bergsonian duration, as in Leibniz's universe

> there are the infinitely small, and perhaps there are only those. Not as quantity but as intention, intensity, because duration is the intention of the mobile and intention is intuition, something that, from the point of view of analysis, always implies a duality, two mathematical points, although in itself it is always unity [...] psychic life is made up of this: a simple intention that is at the same time many intentions [11, p. 277].

Every intensity is, at the same time, simple and multiple. When, however, we replace it with symbols by artificially arresting its dynamism, it always ends up giving itself as one thing or another, one direction or another. Its limpid one-multiplicity is incomprehensible if we approach it with the tools of classical logic because, like infinity according to what Spinoza says about it in the famous letter 32, intensity and duration cannot bear them. Whenever we resort to them to investigate intensity, i.e. divide it, intensity retract annoyed, i.e. undivided. Yet, in doing so, intensity, as well as duration and infinity, favour the triumph of the enemy. As soon as the gaze of the intellect is cast upon the nebula of psychic states, it has the same effect as the Medusa of the myth: an abrupt petrification. Bergson resorts to the image of a sudden drying up of the treasures of the sea that the waves bring on the shore by the impact of the scorching sun; Quantum Mechanics, for its part, uses the concepts of decoherence and desynchronisation of the wave function. But the essence does not change: whenever we attempt to perceive (from *capere*: to grasp, to grasp) what does not seem to be made to show itself, the observed promptly dodges the observer, leaving him with a "ghost of departed life", to paraphrase Berkeley. This, however, does not entitle us to declare the self-contradictory nature of the stream of consciousness proven.

Elevated to the status of object of analysis and goal of a research programme, duration is impossible to rediscover for the same reason that, according to Bataille, interior experience is impossible to consume. Intensity is expressed, but it remains implicit in that in which it is expressed. The difficulty in grasping intensity lies in this: where there is intensity, there we see nothing, if seeing is analysing or knowing in the Kantian sense; when, on the other hand, we see or recognise something, here the intensity has vanished. That is why intensity seems to us to arise from nothing and return to it: while we are grasped by intensity and are insensibly in it, we are not there as subjects, i.e. conscious spectator; once this encounter has "existed" us [46] as such, we will be there instead. As soon as intensity becomes the quality of an object for a subject, and as long as this quality—a conceptual frame of what remains of intensity *post-mortem*—is maintained, intensity is no longer perceptible, because intensity is the being of the sensible, the non-phenomenal trigger of the phenomenon, but the being of the sensible is, at one and the same time, the insensible and that which can only be sensed [24, p. 140]. Insensible from the point of view of representative sensibility—what Bergson calls "intelligence"—intensity is only sensed from that of a visceral sensibility: intuition as *sympatheia* with the thing. It is inevitable, then, that its infinite reality results illogical, the internal relationship that constitutes it contradictory, the qualitative change to which it gives rise inadmissible, and the profound personality with which it coincides unintelligible when we handle it with the tools of the intellect: analysing intensity or duration is like, and the image is still Bergson's, trying to pass a flame with a blade [8, p. 49] or, and it is the same, daring to separate the waters with hands.

Having arrived at this conviction, Bergson abandons the analytic of pure concepts and turns back to its source, the extra-intellectual intuition, taking its data in its entirety as a kind of material whose place nothing else can take. And the first datum of intuition, the authentic immediate datum of consciousness is mobility: not a set of things made, but of things that are making themselves; not a multiplicity of states that are preserved, but of states that change. Every reality, for intuition, is a tendency: a nascent change of direction [10, p. 221]. Yet, only the consciousness we have of our own person, in its continuous making, introduces us into the womb of this restless, incandescent chrysalis. Intelligence, more willingly, goes on the hunt for butterflies by searching the external world for solid footholds to advantageously orient our action in it. Its main function is to find states and entities where there are only processes and fluctuations, to extract "things", like minerals, from that in which they are not found[5]. To this end, intelligence takes snapshots from the undivided mobility of the real, replacing the continuous with the discontinuous, mobility with stasis, changing tendencies with fixed points that freeze their inclination. Intelligence, that is, is not interested in an intimate and absolute knowledge of reality: it only wants to make use of reality. Each of its concepts is a practical question to nature in the terms of which—Kant has shown this definitively—nature will answer. But this answer is miraculous as long as the limits of our perception are taken as the limits of the world and experience is made to match the knowledge we have of it, i.e. as long as intelligence persists in disavowing that mobility has an intelligible existence (movement is the unthought of criticism). The inner dimension of reality, which Kant calls *an sich*, is occupied by the activities that keep it in motion, although what exists for the intelligence is, at most, the result we attribute to mobility after unrolling its loop along the linear space–time drawn by the *Regeln der Verknüpfung*: the laws of nature that only indicate coexistences and successions. Then, while being able to *name* the thickness of reality, critical intelligence is unable to *penetrate* it and settles on surfaces.

6.2.2 *The Imperceptible Small Perceptions*

"Nothing takes place suddenly [...] *nature never makes leaps*" [44, p. 56], but, to the extent that it is pragmatically motivated, our attention to life fixes on signifying units that do not always correspond to the actual moments of the process it experiences. This is why we can so easily believe that reality extends as far as our perception—an anticipation of the existent in the form of a priori construction of its essence—extends and just as easily succeed in making things appear contradictory: the things of our intelligent attention are ghosts and, as such, offer no resistance (impenetrability is the hallmark of the real). Nevertheless,

> there are hundreds of indications leading us to conclude that at every moment there is in us an infinity of perceptions, unaccompanied by·awareness or reflection; that is, of alterations in the soul itself, of which we are unaware because these impressions are either too ·minute and too numerous, or else too unvarying, so that they are not sufficiently distinctive on their own. But when they are combined with others, they do nevertheless have their effect and make themselves felt, at least confusedly, within the whole [44, p. 53].

---

[5] This is the thesis on perception defended by Gaetano Kanisza; but also James Gibson critiques this perceptual model insofar as it is victim of an illusion of embodiment and/or of a tendency towards "corporisation".

Attention requires memory, and when we do not pay attention to any of our present perceptions, we let them pass by without worrying about them. Often, we do not even notice them. If, however, someone calls our drowsy, because selective, interest to mind by pointing out a noise that we have just heard, we instantly remember it and become aware of having had some inkling of it even a moment before, that is, aware of the passive synthesis that, up to that moment, memory has woven for us[6]. Nevertheless, according to Leibniz, we would never be awakened even by the greatest of noises if we did not have some perception of its beginning, which is small, and neither the strongest force in the world would never break a rope unless the least force strained it and stretched it slightly, even though that little lengthening which is produced is imperceptible [44, p. 54]. Yet the continuous character of the passage of nature, as of life processes and psychic phenomena, is incompatible with our logical axioms: its phases coincide neither with complete destructions nor with complete creations while, at both perceptual and intellectual levels, we are presented with only discrete compositions of this kind. The alternative, in both cases, is between perceiving nothing and perceiving something that exists in determined portions, i.e., is already there in sensible amount. This fact, explains James,

> is what in psychology is known as the law of the "threshold". Either your experience is of no content, or it is of a perceptible amount of content or change. Your acquaintance with reality grows literally by buds or drops of perception. Intellectually and on reflection you can divide these into components, but as immediately given, they come totally or not at all [40, p.155].

The difference between intellect and perception, James tells us here, is that the former can tear apart the data of the latter. Consequently, to avoid the logical contradiction of a present infinity that is believed to being traversable in proportion to how well it is divided—*infinitum in actu pertransiri nequit* says scholasticism—

> The quickest way to avoid would seem to be to give up that conception, and to treat real processes of change no longer as being continuous, but as taking place by finite not infinitesimal steps, like the successive drops by which a cask of water is filled, when whole drops fall into it at once or nothing. This is the radically pluralist, empiricist, or perceptualist position [...] We shall have to end by adopting it in principle ourselves, qualifying it so as to fit it closely to perceptual experience. the simplest way would seem to be to abandon continuity and think that change takes place, as perception believes, by absolute integers [40, p.172].[7]

Indeed, one either accepts the logical contradiction, or one admits, and it is James again who confesses this as he plunges one last time into the labyrinth of the continuum in *Some Problems of Philosophy*, that, in passing from one phase to another,

> limit is reached in these successive cases by finite and perceptible units of approach—drops, buds, steps, or whatever we please to term them, of change, coming wholly when they do come, or coming not at all. Such seems to be the nature of concrete experience, which

[6] Husserl's passive syntheses are one of the privileged philosophical terrains in which to see small perceptions at work, those syntheses being, as it were, the product of their activity. For reasons of space, however, we cannot deal with them here.

[7] It is the "drop solution" already anticipated in *A Pluralistic Universe* (Chapter VI).

changes always by sensible amounts, or stays unchanged. The infinite character we find in it is woven into it by our later conception indefinitely repeating the act of subdividing any given amount supposed. The facts do not resist the subsequent conceptual treatment; but we need not believe that the treatment necessarily reproduces the operation by which they were originally brought into existence [40, p. 185].

But, facts, as James himself acknowledges shortly afterwards, do not survive either the conceptual operation, incapable, by its very nature, of adequately reproducing the process that gave rise to it, or the perceptual synthesis. The same thing had been stated, more vigorously, in James' essays of a few years earlier. In the already mentioned pamphlet, for example, James unhesitatingly admits that it is the same nucleus which is able now to make connexion with what goes and again with what comes, as we are sure that the same point can lie on diverse lines that intersect there. Without being one throughout, such a universe is continuous [38, p. 105]. Therefore, given that the intellect thinks of continuity predominantly as a line that tends to infinity and is divisible into infinite parts, and ordinary perception defeats this by relying on the completeness of its data, it must be admitted that

These minute perceptions are more effective in their results than has been recognized. They constitute that *je ne sais quoi,* those flavours, those images of sensible qualities, vivid in the aggregate but confused as to the parts; those impressions which are made on us by the bodies around us and which involve the infinite; that connection that each being has with all the rest of the universe. It can even be said that by virtue of these minute perceptions the present is big with the future and burdened with the past, that all things harmonize—*sympnoia panta,* as·Hippocrates put it—and that eyes as piercing as God's could read in the lowliest substance the universe's whole sequence of events—*Quae sint, quae fuerint, quae mox futura trahantur* [Virgil] [44, p. 55].

Moreover, these small perceptions determine us on many occasions unconsciously[8] contributing to the deconstruction of the fairly widespread belief according to which, *before* any choice is made, we would be, writes Leibniz, in a sort of i n d i f f e r e n c e o f e q u i l i b r i u m [44, p. 56], or neutral state, such that the choice, then, would coincide with an absolute, unprecedented beginning. The free act is a *creatio ex nihilo*, but "ex nihilo", for Leibniz and Bergson, means "from a nothingness of knowledge". *The lex continui* is not violated by freedom when the latter is welded to necessity, i.e. coincides with the *libertas necessitates* defended by the Stoics and their epigones, Spinoza, Nietzsche and Freud. Indeed, even when it comes to the actions, we hold most dear, if we deepen our perception, we always find that we go from the small to the large and vice versa without interruption. The real is narrow, resolute, in the sense that it is decided and constructed by small pulsations rather than by small drops, if by "drops" we mean the big, indivisible and atomic entities anticipated by our perception. Leibniz is clear on this point: a leap from one state to another infinitely different state could not be natural. Perceptions that can be

[8] It is never in vain to recall Freud's warnings, e.g. in the short paper on *A Difficulty of Psychoanalysis* (1919) about the inappropriateness and risk involved in trusting the data of the P–C system in place of doubting the equation psychic = conscious. For the same reason, it is never vain to reiterate that what present itself as *cut* of the conscious weave, i.e. a formation of the unconscious, is a cut, evidently, only for us.

perceived arise in stages from those that are too small to be detected (for Kant the plurality contained in the perceptive unit can only be represented by approximating to negation or zero which, however, is always relative because in its vicinity it only happens that the perception becomes "smaller", infinitesimal—see Kant [41, pp. 290 ff.]. To judge otherwise is not to know sufficiently the immense subtlety of things, which always and everywhere contains a present infinity.

However, it takes that courage of thought [Muth des Denkens] [21, p. 36] that feeds on the source of the continuity of consciousness to judge in this way. The con-conscious transition [39, p. 47]—the relation whereby one experience passes into another into a same self—and, more generally, the continuous transition, have in fact been the *thaumaston* of philosophy since the time of Plato's *Parmenides*. James, before returning to it at the end of his life, had meditated on it at length in his essay *A world of pure experience*, starting from the consideration that

> my experiences and your experiences are "with" each other in various external ways, but mine pass into mine, and yours pass into yours in a way in which yours and mine never pass into one another. Within each of our personal histories, subject, object, interest and purpose are continuous or may be continuous. Personal histories are processes of change in time, and the change itself is one of the things immediately experienced. "Change" in this case means continuous as opposed to discontinuous transition. But continuous transition is one sort of a conjunctive relation; and to be a radical empiricist means to hold fast to this conjunctive relation of all others, for this is the strategic point, the position through which, if a hole be made, all the corruptions of dialectics and all the metaphysical fictions pour into our philosophy. The holding fast to this relation means taking it at its face value, neither less nor more; and to take it at its face value means first of all to take it just as we feel it, and not to confuse ourselves with abstract talk about it, involving words that drive us to invent secondary conceptions in order to neutralize their suggestions and to make our actual experience again seem rationally possible [39, pp. 47–48].

6.2.3 What Science Cannot Do

What we perceive internally in the succession of states of consciousness and what, if we are particularly attentive, we can also observe, albeit exclusively from the outside, in the succession of states of the embryonic development as well as in the succession of states of the becoming flower of the seed—the expression "to be able to hear the grass grow" indicates, in popular language, the pattern of intelligence [21, p. 152]—is that, however different the states are, the transition from one to the other does not stop. Continuity, for the radical empiricism of James and Bergson, is a well-defined type of experience, just as well-defined is the experience of discontinuity that is impossible to avoid when we attempt to make a transition from our own experience to another's, including that of the seed and embryo (for Ruyer, as we shall see, the seed and embryo are also consciousnesses). The intellect, however, thinks of the continuous transition, which for James is the closest relation of conjunction, as a third thing in addition to those from which and towards which the transition takes place: it cannot understand how two things can be distinct precisely at their point of contact. If, therefore, we borrow its logic, we never arrive at a real connection, since "being distinguishable", for the intellect, "means being incapable of connection". We are more successful if we take the reverse path and install ourselves in the mobility of the real by adopting its continually changing direction.

For Bergson, insofar as it is an effort to substitute for the *ready-made* what is in process of *becoming* and to adopt the mobile continuity of the pattern of things [10, p. 224], infinitesimal calculus was born precisely from this inversion. Its intellectual products are truer than those of other sciences because they plunge back into the stream from which they came. Yet, as a science of magnitudes, albeit magnitudes not extensive nor cardinal, differential calculus stops at the drawing of reality. Like the second law of thermodynamics, the new *mathesis* intuits something that is not quantity. But, being a calculus, it is, and remains, a symbolic transcription of what it also grasps, for an instant, outside the symbol. In short: infinitesimal calculus does not think duration within duration: its knowledge is a *knowledge of the interior from the exterior*. The infinitesimal is a notation, albeit a revolutionary one, but a notation is always the transcription of the intention [11, p. 274], and not this intention itself captured as it arises. Halfway between representation and thing, quantity and quality, the infinitesimal is an image, a useful fiction. But even if the image, compared to the concept and the symbol, has the advantage of keeping us in the concrete [7, p. 16], it cannot, as such, replace the intuition of duration for the same reason that we cannot pass from analysis to intuition or from immobility to movement.

The duration that the infinitesimal grasps with the imagination and then lends to the intellect for measurement is that in which we see ourselves act and not that in which we act. Hence, although the philosopher's task is similar to that of the mathematician who determines a function by starting from the differential insofar as both consist of a work of integration [4, p. 185]; see also [34], psychic mathematics is not a mathematical psychology nor, *a fortiori*, a metaphysics: unlike the latter, it proceeds by quantitative rather than qualitative differentiations and integrations, attesting itself, like any other science of symbols, as a spatialisation of duration that inserts intervals where there are none. Despite containing the number in power, a number that is evidently numerative and innumerable (see [51, pp. 12 ff.]) duration in fact does not stop. If it has intervals, insofar as it is nevertheless a continuity of heterogeneity, these are like the indivisible distances of which, according to Deleuze, the intensive *spatium* is woven into its distinction from *res extensa*[9]: something *sui generis* that constitutes an unbreakable intellectual unit [33, p. 256]. Entropy, therefore, is illusory. Even in the most thunderous fall, and even though there is no image drawn from matter, properly so called, that will ever give us the idea of the weight rising [6, p. 267], there is always something working to delay the fall, sometimes even managing to avoid it. Intensity lies motionless on the surface but in its very depth it lives and vibrates [4, p. 204], like the obscene in every scene.

Hence, not aiming at any application, metaphysics can, and indeed must, oscillate between the *intensio* and the intended, the duration and its number, the train and the track to take up the example to which, comparing the work of the scientist and the

[9] On Deleuzian *spatium* and its derivation from Leibnizian space, and on the Leibnizian *spatium* itself, see, among others, [28, 69].

philosopher one last time, Bergson resorts in his dossier on Einstein's relativity [8]. Metaphysics has not to convert its intuitions into symbols. Its task is to penetrate that "absolute position" lodged in the "heart of geometry" by the new maths with an act that, in the bosom of the perceptual datum, makes us foresee something that goes beyond it, as if the perception we have of it suddenly reveals itself to be full of unsuspected and unnoticed details. What it thus loses in usefulness and rigour, it gains in depth and extension because, unlike analysis, intuition is a knowledge *of the inside from the inside* or, one might translate, *of the first person to the first person*.

6.3 The New Way of Thinking: The Radical Empiricism

6.3.1 What Do We Do When We Measure?

The radical, transcendental or superior empiricism of James, Bergson, Whitehead, Ruyer and Deleuze is such because, unlike transcendental idealism, it aims to create concepts that are tailored to the things of which they are concepts, that is, concepts as real and not as possible conditions of experience. Its objects are allergic to the prefabricated and demand the creation of a notion that is their own: not an aprioric possibility that, anticipating them, transcends them, but a strange a-posteriori apriori [14, p. 229] that, by meeting them, identifies them. The concepts of the new empiricism, like its transcendental, are the result of an encounter rather than of a judgement or of a recognition: intensities that are not distinct from the thing they refer to, being, rather, inherent to its *thisness*. Kantian in its aspirations, the new empiricism is more profoundly Humean in dealing with that which the consciousness struggles to recognise and which, therefore, compels it to think. The intelligibility affirmed by transcendental empiricism is that of the aesthetic circumstance: an intelligibility in virtue of which aesthetics itself becomes the jurisprudence, or casuistry, of affectivity, that is a science of occasions in which harmony is made locally instead of being predetermined by eternity. Indeed, that the conditions are tailored to the thing means that is the thing that dictates its conditions in the same way that God once suggested in dreams, to those who took visions seriously, all the information necessary to build a shrine or abbey in his honour. Out of metaphor, this means that a condition tailored to the thing is a condition that, of this thing, expresses the measure, but a measure that the thing itself contains and reveals, under certain circumstances, as its intrinsic nature.

Yet, in the light of the anti-substantialism that characterises radical empiricism, we should say that it is its intrinsic relation, the relation, or bond, in which, as a thing, the thing *qua* thing "dissolves" itself, that the thing reveals, in due course, as its condition. What does it mean, otherwise, that the generation of magnitudes takes place in nature? What if not that nature, but the same applies to duration, pure experience and consciousness, is pregnant with them, and generates them by generating itself in this act? In what sense does the generation of quantities occur in nature if not in that nature itself, in this way, becomes "nature" by making itself available to science, which, therefore, is always science of itself? (see Bitbol's main hypothesis in [13]). The formulas in front of which nature recoils are the formulas that are not its own: keys incompatible with its lock with which, in any case, intellectualism attempts to break into nature's interior in place of letting nature divulge its access codes. That the latter is an interior, moreover, is entirely questionable, given that the psychic, according to Freud [30, p. 300], is extended like Spinoza's *deus sive natura* ("sive" says precisely the diffusion of God in all things) while knowing nothing of this extension with which it scatters in the world, sometimes recovering in bodies, sometimes in notions, just to know itself.

Magnitudes, like bodies, flowers and all figures are born and die in the intemperate time in which a diffuse poignancy, which guarantees their intelligibility, propagates itself on salient forms, discretising itself, one might say, precisely in homage to the indiscretion that wants to extort its secret. And what, even today, questions atomic physicists, no less than phenomenologists and psychoanalysts, is precisely the relationship between this propagation and the gaze that awaits it, and perhaps even cause it: a relationship that, only our myopia makes us believe is consummated between two truly distinct entities, at regular intervals and in protected circumstances. Leibniz's *Suvmpnoia pavnta* states quite the opposite: that each monad observes itself with every other incessantly and the observation is the absolute. The world, in fact, is *omnivoyeur*: a forest of mirrors contemplating which Whitehead concludes that life is robbery [72, p. 105] and every actual entity—its name for "monads"—is somewhere, where "somewhere" means "some actual entity" [72, p. 46].[10] Everything, then, is taken up, reflected, insinuated into every other thing, included and including, implied and implying at the same time. As in the "absolute phenomenality" referred to in the scabrous seventh thesis of the second hypothesis of Plato's *Parmenides*—"if the One is not"—the observer figures simultaneously as part and whole, inner and outer element of its observation. *Suvmpnoia pavnta*, indeed, names an abysmal phenomenality that appears to itself insofar as it appears to someone that appears in it. It is difficult, taking this into account, to shed light on the relationship between observer and observed. Monads are mirrors other than those that allow extensive representation (reflecting mirrors) ([70], p. 66, our. transl): they are formidable multipliers of the luminous experience (resplendent objects) ([70], p. 67, our transl.), living mirrors, absolute icons that perform no iconophoric function.

But then, who mirrors what? Is it an agreement or a disagreement that binds two Leibnizian reflecting surfaces, rather than two Kantian judgments, be they those of thing and its measure, of nature and the Icelander in the famous Leopardi's dialogue, of the electron and the detector in the equally famous double-slit experiment, or of a goddess, Diana, and her, all too curious, observer, Actaeon, in the famous myth? These are capital questions since they concern the fibrous texture of phenomena, the

[10] According to the ontological principle, because of its being *what it is,* every actual entity is also *where it is* [72, p. 59], its actual essence indicating where it is, its situation. The ontological principle is in fact the principle that defines the being of all that is, and being—in this version of the principle—is identical with the having-place of everything.

relationship between a profound legality—that of the passive constitution, of our non-scientific relationship with the world—and the formal, properly scientific structure of mathematics. After Thom's work, we are now quite inclined to recognise that the latter is prepared by the thickening of regularities around one or more attractors that cannot themselves be concepts. Attractors are functions of a modality of experience from which conceptualisations originate and which, thematically, condition the emergence of new ones. Therefore, in the answer to these questions, in addition to the freedom of formal constructions vis-à-vis a silent order that precedes them, the capacity of things to make themselves heard prior to the intervention of intelligence is also at stake.

6.3.2 *The "First Quantification" as the Extension of Intensity or the Becoming Conscious of the Unconscious*

In the chrysalis of pure experience and in the block of marble so often employed by Leibniz to illustrate how the invisible is in the visible, something is announced that the concept shows according to a reciprocal, foundational and non-symmetrical recall, which Criticism, and Newtonian mechanism of which Criticism is a philosophical expression, has never been able to think of, even though it derives from it as a method of investigation. Nor is the contemporary metaphysics of dispositions any more successful because, although it is true that veins, like ideas, are innate as natural inclinations, dispositions, habits or virtualities, and not as actions, these virtualities, even for Leibniz, are always accompanied by some corresponding, often insensible, action. They are virtual, therefore, only from the point of view of what Bergson calls "space" [11, p. 274], due to a retrospective mirage. But retrospection is also an act. Thus, there is nothing virtual in us, nor anything that we do not always apperceive presently, although present apperception, like freedom, has degrees.

Among those who have been most successful in distinguishing them are all the masters of Deleuze and Deleuze himself. Convinced that nature is not bifurcated into the two incommensurable realms of ideas and things, of the intelligible and the sensible, they have managed to solve what for Kant is the fundamental problem—the agreement between our representations and what representation is not—simply by making it the immediate datum: not, once again, something to be arrived at by division, but something from which begin with an asymmetrical synthetic act perpendicular to the schematism of pure concepts. Their aim is to show that we cannot survey degrees of freedom like the degrees of a thermometer: rather, the eye must train itself to endure what Chatelet calls the patience of the oblique [18, p. 87], without ever allowing itself to be fooled by verticality or horizontality alone. All that is contemplated with a purely upward gaze or with an exclusively sideways inspection, destroys the depth because the defect of the immediate data, as of the unconscious, is of extension rather than nature. Knowing them from the inside, therefore, only requires expanding the boundaries of perception to the point where it can see, with undoubtedly bastard reasoning in the sense that Plato gives to "bastard" in his *Timaeus* (Tim., 52 b2) that analysis and measurement begin when the tension of duration vanishes and the heterogeneity of the qualities which succeed each other in our concrete perception is sufficiently diluted to become, from our point of view, practically negligible [4, p. 238][11].

For Bergson, quantification automatically becomes possible when the turbulences of perception stabilise and the differences among intensities diminish in the *momentum* in which duration's energy wanes until it almost disappears. Thus, *mathematics begins exactly where the psychic ends*, almost as if it took over from it in the same way that consciousness, for Freud, arises instead of a memory-trace [29, p. 19]. Precisely for this reason, despite the fact that it seems to us that intelligence abusively takes advantage of the subdued nature of sensory impressions by constructing on the remains of duration a presence that at least most scientists hope will avert its return, we must recognise that it is duration itself, which Bergson compares, not by chance, to an elastic band, that contributes to the formation of the blade with which intelligence will later vivisect its undivided act. In other words, it is one and the same movement by which intuition intellectualises itself by determining itself into distinct concepts and what we call "matter"—the perceptual *positum* or mobile point that moves from one instant of space–time to another—fragments into *extra partes*. This means that, if a simple shift of the matter in space can summarise the universal transformation [4, p. 277], it is because it is the same duration, the same creative impetus that Bergson calls "élan" and is infused in all things that spontaneously stretches out as if it were seized by fatigue due to the excessive concentration that its mode of generation entails. The slightest homogeneity detected in an ocean of heterogenesis, in this sense, is already all the a priori space of the phases which is readily spread out under the unfolding of the duration like a shroud, and if the qualified *extensa* seem to be the only reality even though, of the only reality, they are ectoplasmic doubles, is because, with a motion as suicidal as it is vital, intensity buries itself in the products it gives birth to and which, for this reason alone, obscure it. Its fate, like that of the unconscious saying that remains forgotten behind what is said in what is meant, seems to be death in childbirth since, even where the simultaneous genesis of intelligence and matter is concerned, the intuitions underlying the paradoxical phenomenology of intensity sketched by Deleuze in the fifth chapter of *Difference and Repetition* remain valid.

The latter is a phenomenology without phenomena, very similar, although we cannot deal with it here, to the phenomenology of the inapparent patiently reconstructed by Freud in *Psychopathology of Everyday Life* because even the conscious system is characterized by the peculiarity that in it, in contrast to what happens in the other psychical systems, excitatory processes do not leave behind any permanent change in its elements but expire, as it were, in the phenomenon of becoming

[11] In Lectures at the Lycée Henri IV we read that the ultimate elements of matter, although simple substances like ours, are less rich in changes and properties of all kinds. They are not unconscious, perhaps, but of a consciousness so obscure and slow as to be *practically* equivalent to unconsciousness [5, p. 59].

conscious [29, p. 19]. This means that both the extension of the intense as well as the conscious becoming and the spatialisation of duration are transformations due to a concomitant substitution of a suspension [6, p. 259. The physical, for Bergson, is simply psychics inverted (Ibid., p. 221) and we pass from the first to the second by way of inversion, or perhaps even by simple interruption (Ibid., p. 220). There is a change of set-up, something akin to a phase transition. The alternative between space and duration, extension and intensity, real predicate and being, to use the Kantian pair, is so clear-cut that all that which seems positive to the physicist and to the geometrician would become, from this new point of view, an interruption or inversion of the true positivity, which would have to be defined in psychological terms (Ibid., p. 228).

The task of metaphysics, after all, is precisely to construct a cosmology that would be, as it were, an inverted psychology and even for Ruyer what is true geometrically, is false psychologically. However, if intensity, like duration and the unconscious, eludes the present to the extent that it has a degree only in function of its cancellation, what is it that flashes and suddenly appears? The unconscious, consciousness or that *monstrum* Freud calls "unconscious consciousness"? Intensity, extended or intensive magnitude, that illegitimate translation of duration into space, which Bergson, in his *Essay on the Immediate Data of Consciousness*, dismisses as an abusive mixture of what should remain strictly separate: intensity and magnitude, psychic and physical?

6.3.3 The Horizon of Events

For Bergson, intelligence has fabricated the hybrid notion of "intensive magnitude", like the other two discussed in the *Essay*—"number" and "spatialised time"—by surreptitiously introducing the homogeneous into the heterogeneous when there is no point of contact between the unextended and the extended, quality and quantity [3, p. 70] considered in themselves. And no matter how much the juvenile dualism is diluted in later works: the "substitution via interruption" paradigm that rages in *Creative Evolution* seems to rehabilitate it under the surface. For what is an inversion and what do we observe if the exhaustion *of intensio* is the price *intensio* itself pays for its publicity, if, that is, the loss of self is a condition of self-science? How can being arise immediately and in finite quantities if quantity seals, on balance, its non-being?

The proportionality between Thom's saliences and pregnancies is inverse: a plus of notation corresponds to a minus of intention, and when the latter intensifies with new sap, the former reveals itself to be inadequate, enjoining a rewriting: it is the aleatory, or the sublime, the crisis of presence-consciousness as a catastrophe of the schematism that organises it due to the sudden emergence of rhythm—what Kant, in the *Analytic of the sublime*, calls "aesthetic dimension of magnitude"—from the depths of the soul. But what is a crisis? And in what sense can we speak of a suddenness? How does it happen that something signals itself and imposes itself on our attention if this something is felt in a sensation that, Kant also recognises [41, p. 291], is neither spatial nor temporal despite being, as a sensation, a modification of our spatial and temporal sensibility? How can this something, which Kant also calls "degree" or "intensive magnitude", pierce the diagram of *Transcendental Aesthetics* without piercing consciousness? How it can stand out, or simply begin to be, in the continuous unity that characterises our faculty of receiving, i.e. our *Sinnlichkeit*? How, if not by imagining a deeper alliance, a pact between inside and outside drawn up, one might say, in the absence of the contracting parties?

In a text of a few years ago, Sarti and Bifo define the aleatory as follows: an incompressible sequence, irreducible to algorithms capable of generating it that are shorter than it; a sequence, therefore, that must be reproduced entirely as it is if one wants to describe it [63, p. 100]. The sequence in question, we can translate, is the fundamental measure of the phenomenon perceived, precisely, as aleatory, i.e. the germinal relation that generates it and can be found, for example, in the works of Peirce, Simondon and Deleuze. Everything that appears is, in fact, a sign. Every phenomenon is a sign that refers to an inequality, or indivisible distance, as its condition. But this original difference

> does not confine itself inside contradictions; on the contrary, it opens out and, like a spore case inflated with spores, releases them as so many singularities that it had improperly shut off, some of which he intended to exclude while retaining others, but which now become points-signs (points-*signes),* all affirmed by their new distance. The disjunction, being now inclusive, does not closet itself inside its own terms. On the contrary it is nonrestrictive [26, p. 77].

Inclusive and incompressible at the same time, the sequence is also illimitable. It gives itself, first as enveloped, as pure similarity: *icona sui.* Then, it allows the difference between the term and its own displacement to explode. The categorial primacy to which the germinal relation corresponds designates in fact the way in which something would be in itself, independent of anything else, so that it would make no difference if nothing else existed [52, p. 7.538]. This is why it can be considered entirely for itself. But since its state is what Simondon calls "metastable", its firstness can also displace itself in relation to itself by surveying on either side of its centre (however, this displacement is immediate because near a bifurcation the transmission is spontaneous—Simondon 1958, trans. 2020, p. 65). What seems to us to be a relation thus turns out to be a a splitting, a resolution, a non- symmetrical distribution occurring in a totality starting from a singularity (Ibid., p. 51) because the metastable state is a state of *distinction without discernibility* such that singularities are progressively identified in the form of parts external to one another—living organisms, phonemes, magnitudes. According to Simondon, we can speak of diversity and analogies only with regard to these "parts". Therefore, if the sequence cannot be manipulated at will, nor can it be anticipated by perception, it is because what is manipulated and perceived are its products or, in Bergson's language [11, p. 281], the curve manifested in space. In the vibrating and generative tangent that, in the meantime, continually stands out in duration like the spindle *of Ananke* consecrated by Plato in the myth of Er, rather, we stumble upon and, when this happens, what we can do is only to take note of it or, as it is said in the passage by Sarti and Bifo just quoted, "reproduce it entirely as it is". In a word: mimic it.

There is, however, a third option. And it is Bergson himself who formulates it in the 1989 *Essay* precisely to escape, it must be said, from the *fork-model* gallows with which we usually think of freedom (see [3, p. 176 ff.]). Freedom, for Bergson, is not a choice between two equivalent states available to our evaluating gaze, it is not the Christian *libertas arbitri*, nor the Cartesian *libertas indifferentiae*. Freedom, for Bergson, rhymes with authoritativeness, self-expression. And the self, the profound self, indiscernible from duration, is not something we choose but something we eventually liberate, thus liberating ourselves in turn. The self is destined. Therefore, just as indeterminism does not explain freedom, neither can vulgar determinism. Bergson is categorical about this: the only way to predict what x will do is to be x, to become x, i.e. to intuit it, to enter it [3, pp. 184 ff.]. Not "to make a tracing of it" but, to quote the mediating image of *Rhizome*, the preface to *Mille piani*, "to make a map out of it" [27, pp. 12 ff.], as modern mathematicians have done with motion, considering it the motion of an agent, and as Einstein did by installing himself on the horizon of speeds.

Perched on a photon [50, Chap. 3/4], the father of Relativity rediscovered the scholastic tradition that first posited a maximal degree of perfection and then degraded it towards the infinitely obscure, the degree zero, and, at the same time, exceeded it. Speeds, in fact, must be contemplated solidly in a spectrum like the different distance lines of a painting. But the compact intuition that authorises their putting into perspective cannot be achieved without a decisive transport to the fan-point where the velocities open: the horizon. As the Renaissance masters of perspective, and Dante before them, intuited, there is a place in which, bound with love in a volume, that which is squared off for the universe is internalised (PD XXXIII 87, our transl.): the central axis of a game of balance between the vertical and the horizontal that takes them all on. But degrees are not confused there. The horizon, or degree zero, is not a common denominator but a joint that discerns the degrees by unfolding them with the same dexterity with which an embryo unfolds its totipotence. The compactness of the similarities that is consummated in it distinguishes them more than their presentation one by one, because the metric element, properly speaking, is the tribute of a neutral observer in charge of measuring the intervals between things, the distances from point to point. As such, it can only express the referee of a combat already won at the start by the spatial element. And it is to this abdication of the measurable that Einstein returns to oppose the power of the horizon: the projective element that initiates thought into an oscillation between all finite degrees of distancing.

In the horizon, the finite is reunited with the infinite. Then, its absence destroys the creative homogeneity—the pure similarity of the germinal relation—provided, instead, by depth. This homogeneity does not result from the extraction of an average. Averaging ratifies *après-coup* the appearance of a collection of objects by producing a sample-unit whose iteration exhausts it extensively. But averaging is only the operation required by a finitude that has divorced itself from the infinite which demands a homogeneity that does away with relief—even if it is the derisory relief of a panoply of differences—and that, over time, levels everything.

6.3.4 *Becoming Familiar with Things, or how to Know their Inside from Inside*

Galileo put himself in the place of nature which, in its simplicity, could not possibly have chosen to make an immense number of bodies move with inestimable speed in order to achieve what could have been achieved by the mediocre movement of a single body around its own centre, for the same reason that Einstein teleported himself into the photon and Archimedes, centuries earlier, had immersed himself in the vat imagining that his body was nothing more than a bag of water: to understand the photon, one must make oneself a photon, in the same way that, to understand what floats, one must make oneself a wineskin. Bergson says no more when, in *The Perception of Change,* he affirms that the only way to know whether the changes that take place outside of us are a single movement or composed of multiple movements between which stops are interspersed, is to make oneself interior to beings and things, as we are interior to ourselves [10, p. 189]. This becoming interior to which, still in *Duration and Simultaneity*, he appeals as the only solution to resolve the paradox of the twins (see [8, p. 9 ff.]), is a metamorphic, or diagrammatic experience, in which, by withdrawing from the world of beings, *thought becomes nature with the same act in which it becomes intelligible, that is, thought.*

The diagram is a "chaos-germ" [25, p. 102]: an experience and a self-experience at the same time. The gestures it captures, and especially those it arouses, are no longer directed towards the things we see, but participate in a descent of diagrams, of a technical becoming [18, p. 36]. Yet this articulation between the figural, the ideal and the operative is not only crucial at every constitutive level of scientific practice. Even in metaphysics as integral experience [10, p. 236] there is always an interweaving of understanding, intuition and action because, as a painful effort to go beyond the human condition [7, p. 77], metaphysics too places itself on the horizon of events to follow their combination from close. In other words, the horizon also provokes and governs that experience of thought which occurs when, by means of intuition one places oneself at the *turn* of experience, profiting by the faint light which, illuminating the passage from the *immediate* to the *useful*, marks the dawn of our human experience, there still remains to be reconstituted, with the infinitely small elements which we thus perceive of the real curve, the curve itself, stretching out into the darkness behind them [4, p. 241]. This means, as Baumgarten says in his *Aesthetica*, taking care of confusion [2, p. 15] by bringing knowledge to its perfection.

The horizon, like Plato's *exaiphnes* and Bruno's atom, is immense, articulated, as an intense complex or an extended simple. But it is not confused. Therefore, insofar as it is also an "offensive" philosophy, in the sense Chatelet gives to the adjective (see [19]), metaphysics must sit at the outposts of the resolutely obscure. The latter is not diabolical or refractory to logic. It is, rather, that through which the advent of new logic as well as new dimensions can occur. An offensive philosophy, that is, must make the negligible relevant because "intensity", "duration", "pure experience" are not just names for the immediate stream of life that we feel flowing as soon as we equip ourselves with a wide-ranging, background sensibility. Or better, they are names for this stream, yes, but the stream, like chaos and disorder, is not formless.[12] That even the obscure is distinct means that the obscure provides the entire material of our subsequent reflection in the form of of a that which is not yet any definite what, though ready to be all sorts of whats; full both of oneness and of manyness, but in respects that don't appear; changing throughout, yet so confusedly that its phases interpenetrate and no points, either of distinction or of identity, can be caught [39, p. 94].

6.4 The New Way of Thinking: Ruyer's Psychobiology

6.4.1 The Need for a Further Paradigm Shift

Pure experience flows through adjectives, nouns, prepositions and conjunctions because its purity only has a relative meaning: it means that intensity "that has returned to dig in the shallows" [24, p. 197]. As far as we can go back, then, we find that the flux, both as a whole and in its parts, is made up of conjunct and separate things. Indeed, if we grasp the inside from the inside, what we experience is that term and relation, disjunction and conjunction, primary and secondary qualities are equally real, equally primordial, and that the flow immediately tends to fill itself with relevant elements which, as salient parts, will then be identified, fixed and abstracted by the intellect. It is in fact

> the same act by which I feel that this passing minute is a new pulse of my life, I feel that the old life continues into it, and the feeling of continuance in no wise jars upon the simultaneous feeling of a novelty. They, too, compenetrate harmoniously. Prepositions, copulas, and conjunctions, 'is,' 'isn't,' 'then,' 'before,' 'in,' 'on,' 'beside,' 'between,' 'next,' 'like,' 'unlike,' 'as,' 'but,' flower out of the stream of pure experience, the stream of concretes or the sensational stream, as naturally as nouns and adjectives do, and they melt into it again as fluidly when we apply them to a new portion of the stream [39, p. 95].

However, what leads us to call "confusion" the real movement is the fact that it presents, as if dissolved in one another, a multiplicity of different aspects of which conceptualisation, by keeping them apart, violates complicity. But are the differences so dissolved in each other? Does not each moment of experience have its own quality, its own duration, its own extension, its own intensity, its own necessity, its own clarity, and many other aspects, none of which can exist in the isolation to which our discursive logic condemns it? Although in duration we only encounter examples of the same in a crossroads of differences that exist as if dissolved in the same thing, *qualia* differ. Moreover, Leibniz had already welded the *lex continui* to the principle of the identity of indiscernibles admitting a real plurality of substances. Like the law, the principle is also based on the actual existence of small perceptions because it

[12] Husserl's passive syntheses arrive at the same results: sensible experience must have within itself the criterion of its own meaningfulness as well as intuition must independently produce its own formal content.

is by virtue of insensible variations that two individual things cannot be perfectly similar and must always differ more than by number.

Already in Leibniz, therefore, we find the members of the endiad that titles this volume, "plasticity" and "discontinuity", together with the poles that characterise the field of differential heterogenesis, "iteration" and "variety". Defined as the emergence of singular forms from assemblages of heterogeneous operators [64, p. 4], heterogenesis is in fact at once repetition and difference, replication and specification, multiplicity and unity. There are ruptures, alterations, novelties, but ruptures in the link, alterations in the preservation, novelties in sameness. The world of heterogenesis, like that of pure experience, is a mosaic whose tesserae are not held together by their bedding, for which bedding the Substances, transcendental Egos, or Absolutes of other philosophies may be taken to stand. In radical empiricism there is no bedding; it is as if the pieces clung together by their edges, the transitions experienced between them forming their cement [39, p. 86]. But why, then, do neither Leibniz nor his invention, monadology, appear among the first-order theoretical references of the "Dada turn" in mathematics?

Deleuze's name stands out in almost all writings on the new mathematics, but only rarely, indeed almost never, do we encounter those of James, Bergson, Whitehead and Ruyer. This is surprising because, in addition to having been inventors of new and revised monadologies—cosmologies obtained by overturning psychology, i.e. by asking intuition to model ontology [55, p. 346]—the metaphysics fielded by the radical empiricists, at least on a philosophical level, could put further tools in the toolbox of heterogenesis, perhaps even provide it with new boxes in the form of conceptual frames more suited to grasping what really interests it—the absoluteness of the passage—and more suited to allowing its mathematics to correspond to it. Provided, however, that one accepts the impossible: the psychic becoming of mathematics or the mystical becoming of mechanics, if we consider the appeal addressed to humanity by Bergson in *The Two Sources of moral and religion* ([9], trans. 1974, pp. 255ff). After all, isn't this what is required of us by what can only with difficulty be made to fall into the ranks of the scientific theories known to date, namely Quantum Mechanics, given that it is, rather, another science, another way of doing science?

Now, the paradoxes of Quantum Mechanics are the same paradoxes that Ruyer understood as paradoxes of consciousness in the work of the same name [61] and it is by changing the paradigm that they are most likely to be resolved to the benefit of the Platonic realism of absolute forms—the simplest constituents of nature according to Ruyer—rather than of Einstenian realism: a local realism naively faithful to the principle of causal contiguity and Whitehead's simple location. Indeed, even if our units of measurement are conventional and foreign to nature's intentions, given that none of our mathematical symbols can express the fact that it is the body in motion that is in motion rather than the axes or points to which it refers because true mobility, as such, eludes scientific knowledge, the abstract succession of positions that retrospectively marks the undivided growth of duration is contained in the nascent state of duration itself. The latter, at the height of its contraction, is pure intensity: a tension towards extension. But extension is a tension which is interrupted ([6], trans. 1944, p. 267).

Something abrupt takes place, both in the sense that it happens and in the sense that it occupies space, generating it spontaneously by its own happening ("spontaneous" from the root "span" has the sense of spreading out, moving towards, having space, extending itself, pushing, succeeding) as a result of a weakening (morphogenesis has a catastrophic character—[63, p. 132]. The primacy, here, seems to be of the fall. Yet, everything that falls is curiously active.

The active is the fall [25, p. 80] says Deleuze in his work on Bacon, because it is a single volume that first contracts and then stretches until it fades into a lifeless aggregate. For lack of a better word, but very significantly, Bergson calls "consciousness" the principle that has only to let go its tension, may we say to *detend*, in order to *extend*, the interruption of the cause being, here, equivalent to a reversal of the effect [6, p. 259]. Yet, he takes care to specify, immediately afterwards, that this consciousness is not the narrowed consciousness that functions in each of us [6, p. 259] and that, if the same kind of action is going on everywhere, whether it is that which is unmaking itself or whether it is that which is striving to remake itself, the image of a centre from which worlds shoot out like rockets in a fire-works display (ibid., p. 271) is particularly adequate to express both this similarity and that consciousness, provided, of course, that the centre is considered as a continuity of shooting out (ibid., p. 271) rather than as a thing. Thus conceived, says Bergson, creation is not a mystery: we experience it in ourselves when we act freely (ibid., p. 271). It becomes obscure if we think there are created objects on one side and a subject that creates them on the other. It ceases to be so as soon as intuition, there where the intellect, exercising itself on the supposedly fixed image of the unfolding action, shows an infinite multiplicity of unconnected parts, discovers a simple process: an action which is making itself across an action of the same kind which is unmaking itself, like the fiery path torn by the last rocket of a fireworks display through the black cinders of the spent rockets that are falling dead (ibid., p. 264).

6.4.2 The Primacy of the Psichic

Duration is differential heterogenesis, qualitative change, a succession of sudden condensations of recognisable forms by virtue of the decay of eternal information in ordinary space–time [63, p. 109],[13] i.e. of the *élan*, the germinal relation of the cosmos. However, if the avant-garde of mathematics wants to remain faithful to the embryogenesis of the world, to quote the title of Ruyer's last work [62], and to the perennially becoming character of nature,if it wants, that is, to go back to the shoulders of the structure and not limit itself, as happened with dynamic structuralism, to modifying only some of its components, it must call the principle of the phenomena that interest it "consciousness", training itself to reconvert its symbols into the intuitions from which they spring. The function of metaphysics, according to Ruyer,

[13] From this point of view, *heterogenesis* is a *heteriumena*: a composition of differentials on the virtual plane; a condensation or actualisation of singularities.

is to transform scientific observations into a knowledge of bonds and senses [58, p. 224], i.e. into a knowledge of consciousness. Yet even a mathematics of creative duration or cerebral plasticity, as differential heterogenesis would like to be, must train itself for this transformation by placing itself from the point of view of the inner making of things. Differential heterogenesis too, that is, must make use of science to interpret psychological intuition and of this intuition to imagine what lies behind the symbols if it wants to find subjects behind its "heterogeneous operators"[14] by striving to imitate them not in their results but in their procedures. Indeed, that the laws of motion cannot derive only from extensive or impenetrable considerations, means that there is something else in the bodies from whose modification impulse and impetus result.

For Leibniz, if monads did not exist, everything would be appearance. If we consider the spiritual element as the exception rather than the rule, the book of nature, even for Fechner, becomes a volume of mechanics. It matters little, from this point of view, that for a long time it was believed that the monad, like the differential, was only a hazardous abstraction: it is the sensible finite, today, that is considered a hazardous *concretum* [ein g e w a g t es Co n c r e tu m] [21, p. 145],[15] a poorly analysed mixture, supported, at most, by an imaginary and substantialist conception of number and dogmatic of matter. The latter, for Leibniz, is nothing other than the mixture of the effects of the infinite that surrounds us: what Spinoza, in his *Ethics*, calls *facies totius universi* and introduces as the only example of a mediated infinite mode. Thus, nothing but an image. Whereas both the monad and the differential are real *mondes pleins d'infintié*: living principles of sensible nature and finite quantities.

Admitting this means recognising, with Deleuze, that intensity is always the answer to the question "who?" [24, p. 274] and disposing oneself to intercept, at every stage of every dynamic, the intention on which this who depends, and that it is not a thing but a form, a substance-function: a pure bonding activity. Elsewhere, by replacing the atom of matter with a quantum of action, Quantum Physics has long since paved the way for us by demonstrating that the being of the atom is indiscernible from its behaviour. Its findings are crucial because they reveal that particles are less things than activities, zones of forces sustaining a given form, and what Ruyer calls "primary consciousness" is precisely the way these zones are coordinated so that a given form is maintained. With one, but not irrelevant, difference: primary consciousness is, together, this same shared form. That is, like Leibniz's monad, vision and

[14] These are qualitatively different bonds for each point and phase, created by becoming itself, i.e. local absolutes. But between these constraints and the primary bonds improvised by primary consciousness, are we sure it is possible to find such a great difference that it is not possible to reduce the former to the latter? Indeed, if we take the definition of heterogenesis as «the emergence of singular forms from assemblages of heterogeneous operators» (Sarti/Citti 2020: 4), we can see that the emerging forms are the new bonds, or degrees, of the process; that the assemblages from which they emerge are very similar to consciousness as a field of non-local intensive bonds—a rhizome, a smooth space, to use the terms of Deleuze & Guattari's *A Thousand Plateaus* [27, pp. 474 ff.]—and that the operators are precisely the elements that make up consciousness as a *unitas multiplex*: an individual of individuals or form of forms.

[15] On the rehabilitation of panpsychism see the work of Philip Goff (see [31, 32]).

force at the same time; or to use Ruyer's terms, knowledge and being, "idea-form" and "idea-structure" together; the diagram, we can also say, and the cuttlefish bones that, from time to time, remain on the beach.

Ruyer's primary consciousness is a self-forming form that sustains its organisational configuration through time[16]: a self-sufficient activity of creating and maintaining bonds in which there is no distinction between being and acting. The organic world simply manifests the same great fact: form is inseparable from matter. Living matter only ever appears as formed [60, p. 31]. And Quantum Physics is important because it proves this at the level of the ultimate constituents of what, we should by now say, is only a supposed "matter". In the atomic world, there is no distinction between force of consciousness and physical force: an atom is, together, the tailor and the cloth he shapes, the painter and his canvas. Its primary consciousness is a strange code that writes and reads itself. Hence, what appears to the physicist as a link through an energy exchange is only an elementary field of consciousness [60, p. 193], in the same way that equipotentiality is the objective functional aspect that a particular mode of reality assumes for an observer: a consciousness—that is [...] an absolute form, an absolute self—surveying domain [58, p. 74]. Not, therefore, a magical property of material tissues and their chemistry (Ibid., p. 56). Ruyer considers the attribution of an occult *qualitas* to matter to be the central problem of traditional vitalisms. For him, there is neither a conscious force separate from that on which it exerts itself nor, consequently, an inert matter: all matter is living.

6.4.3 Two Kinds of Reality, Two Modes of Knowledge

According to Ruyer, consciousness is the subjective counterpart of what, from the outside, appear to us as material phenomena[17] and what we call "material" as opposed to "living" is merely the figure isolated from the background that constitutes and besets it. Just as Bergson distinguishes between two species of multiplicity, and today we distinguish between the organic and the inorganic, Ruyer distinguishes between absolute forms—real beings—and molar structures—statistical and secondary compounds fabricated from the former. The only difference he admits is that between the psychic and the mechanical, individuals and crowds. Examples of the former are molecules, viruses, embryos, atoms and organisms; examples of the latter, clouds and gases, as compounds of individual molecules, calcareous sedimentary formations as aggregates of individual molluscs or, again, masses of human beings as collections of individual consciousnesses. As we can see, it is a distinction that owes much to Leibniz and concerns the capacity for self-formation of which molar structures, as mere discrete multiplicities of juxtaposition, are devoid: only

[16] But the monadological inspiration also came to him from the microphysics of a work that Ruyer knew well: *L'Homme microscopique: essai de monadologie* (Augier 1952) which opens up to a "monadologie Générale".

[17] And so, after an initial phase of adherence to the mechanism, Ruyer moves on to pampsychism.

absolute forms self-organise themselves by mutating and at the same time sustaining the morphogenetic process of creative differentiation. But it is a distinction that also entails a new distribution of the sciences: the primary sciences are those that focus on absolute forms; the secondary sciences are those that study individuals from the molar or statistical side, like most of the traditional sciences that have been privileged until now: from classical physics, e.g. thermodynamics, to physico-chemical physiology, via population biology, political economy, genetics and neuropsychology. The latter, for Ruyer, fall victim to what Whitehead has called the "fallacy of misplaced concreteness" because, observed from above, the movement of a crowd of human beings during a demonstration is indeed explicable by the laws of fluid dynamics, but this dynamic says nothing about the individual subjects who set it up and participate in it. For the same reason, a calcareous sedimentary layer can indeed be described in terms of its physical and geological properties, but it cannot be confused with the individual molluscs that make it up.

From molar and statistical properties to individual ones, as from analysis to intuition according to Bergson, *non valet consequentia*: subjected to the patience of the oblique, these properties, together with the units of measurement that express them mathematically, coincide again with the multiplicity of movements that take place within the chrysalis of which they are but the metamorphosis. This is why, without contradiction, an elephant can be defined as a molecular and microscopic being, much more so than, say, a soap bubble [58, p. 209] and a cell can be estimated to be more powerful than a volcano. The data of perception, even scientific perception, are not simple, and Ruyer's merit is to have claimed, in the light of this complexity, the priority of the sciences that follow the infinitely small lines of fibrous continuity between absolute forms, such as Quantum Physics, biology and embryology, psychology and sociology. Similarly to what Valéry did in front of the most eminent representatives of aesthetics in 1937, Ruyer too, thinking about the new difficulties of Quantum Physics and all the uncertain discoveries the latter is forced to make and rework every day, wondered whether, after all, the retina could not also have its own views on photons, and its own theory of light, if the corpuscles of touch and the marvellous properties of the muscle fibre and its innervation could not be very important stakeholders in the process of building time, space and matter. Physics should return to the study of sensation and its organs ([68, p. 19], our transl.),[18] i.e. become psychobiology[19]: only in this way can artificial conceptions dissolve and re-tune themselves with duration. Indeed, however much the eternal information that Bergson calls *élan* decays, even in a quantum sense, every time

[18] But Bergson, too, had found himself supposing that "the ultimate elements of matter are not without analogy with consciousness: they are, to tell the truth, poorer or slower consciousnesses, but they nevertheless have continuity and simplicity [...] It is not difficult to find in the relations that human intelligences entertain among themselves an image of the relations that things in themselves entertain among themselves [5, p. 59]: things in themselves, precisely because they are in themselves, are intelligences.

[19] As is well known, Penrose, in his works, admitted several times the existence of a "quantum sensitivity" which is very similar to the photosensitivity that Bergson attributes to matter in the first chapter of *Matter and Memory*.

it is subjected to conceptual analysis—a bombardment of photons—it increases as soon as we abstract, so to speak, from our abstractions to return to the undivided life from which they come. *Élan,* in fact, is the cause of all the tensions and relaxations, impulses and inhibitions of this life, and remains even when *substracto autem motu ab actione et passione*, as an unbreakable intellectual unit: that dominant chord that Mozart so often hastened to put down again.

Therefore, it is only by presupposing this sonata that is consciousness that the problems that confound the intelligence before the prodigy of equipotentiality, *entanglement*, or superposition of states, and wave-particle dualism find a solution, the first being the objective functional aspect that takes, for an observer, consciousness as an absolute subjective form, the second by exemplifying the invisible and non-local link, an asymptotic synthesis or domain of possible delocalisation of what will be observed as corpuscle, that characterises this consciousness no less than the plexus of qualitative *nuances* squeezed together in the distance of an intensive magnitude, and the third, finally, by being the result of the tautohegoretic nature of consciousness in virtue of which it is, at once, subject and object of itself, *forma et materia sui.* Distributing this consciousness in the infinitely small, assuming that panpsychism is the truth of monadology [59, pp. 33–34], what appears as aleatoriness, unpredictability or indeterminacy to "knowledge-observation", to "knowledge-participation"—to use Ruyer's terms—appears instead as the sign of a very clear determination and decision. These phenomena, indeed, indicate, perceptibly, the insensible presence of subjects, none of which, however, can be observed from the outside.

In other words, impenetrability is not, contrary to what was believed up to Kant and perhaps beyond, an intrinsic quality of *res extensa*,[20] but the mark of a creative duration, in the same way that the solidity, of wood, metal or any other body, is not the symptom of its rigidity, but the manifest emblem of the deeper solidarity that binds its elements. And we must get to this point where the elements are immediately unified [58, p. 105] through unobservable non-local bonds: only once here does what from the outside appears to be a probability turn out to be a destiny. Ruyer says that what, on the outside, looks like probability, on the inside is a becoming and that primary consciousness is this becoming. But there is no difference: destiny is, first and foremost, the unity of various physical causes with each other, and this unity does not cease to make itself for the same reason that, as long as we are alive, destiny does not cease to fulfil itself.

6.4.4 The Absolute Domain of Survey

Following both Bergson's remarks on the nature of duration and those of Ruyer on the nature of consciousness we can see that the *loop* in which the life of a spiritual automaton is resolved contains ("enveloppe", as we saw, is the verb chose by

[20] Classically understood impenetrability is the quality that objectifies a tactile perception. But touch only registers superficial interactions.

Chatelet) a certain number of presents succeed each other in such a way that, no matter how strong the inconsistency or possible opposition between them is, we have the impression that each one represents the same life at a different level, composing, precisely, what is called "a destiny". The latter, Deleuze explains,

> never consists in step-by-step deterministic relations between presents which succeed one another according to the order of a represented time. Rather, it implies between successive presents non-localisable connections, actions at a distance, systems of replay, resonance and echoes, objective chances, signs, signals and roles which transcend spatial locations and temporal successions. We say of successive presents which express a destiny that they always play out the same thing, the same story, but at different levels: here more or less relaxed, there more or less contracted. This is why destiny accords so badly with determinism but so well with freedom: freedom lies in choosing the levels [24, pp. 102–103].

The succession of present presents is nothing more than the manifestation of something of the way in which each present resumes throughout life, but at a different level or degree from the previous one, all coexisting levels or degrees that present themselves for our choice on the basis of a past which was never present [24, p. 103]. And since primary consciousness is this concealed art of coordination and selection that unifies *in fieri* a multiplicity of phases, what the gap between the real beginning and the phenomenal beginning of movement gives us is not the illusion of the visionary nature of the machine, as Sarti and Bifo argue in their essay (see [63, p. 39], but the certainty of the activity, this one visionary, of a consciousness that devises each type of machine as a cause without a place or function without an organ.

For Ruyer, primary consciousness is in fact a formative activity [60, p. 39, 47, 94–95] hat invents its own act and itself in the act, just as the protozoon improvises pseudopods made of its own protoplasm with a view to its locomotion, and man puts one foot in front of the other to walk [60, p. 6]. Consequently, if we want to account for the complex unity of every morphogenetic process, it is from this fundamental property of being that we must start. Contrary to what has been argued by cybernetics and relational structuralism, including dynamic[21], Ruyer is convinced that chaining or assemblage is not the primary datum. Consciousness does not emerge from a relation as a quality irreducible to its premises: it is that quality and that relation insofar as it is nothing more than binding force [force de liaison] [58, p. 107]. Then, strictly speaking, it neither emerges nor is invented. Consciousness preserves, lasts, and it does so by means of those organs that, on a global scale, are the organisms created by itself as *pied à terre* to which it is conjoined according to operation but not, to take up an ancient distinction, according to being. As a form, that is, consciousness is both a dynamic principle, a mechanical structure and a perceived figure, and it is

[21] According to Sarti and Citti, in their work Thom and Petitot have varied the actual but not the virtual. The latter remains homogeneous in the form of rigid constraints. In the same way, the paradigm of 'phase transition by symmetry breaking' is insufficient as long as it is not the equation, i.e. the virtual, conditions that change from time to time. This is why it is not enough to make the structure dynamic, but we must strive to see it emerge, and to see it emerge means 'to enter into it'. The virtual varies as the Bergsonian conscience *se faisante* varies. For heterogenesis, primacy belongs to the becoming and to the conditions that the becoming generates by becoming, that is, by adjusting and correcting itself. These emerge as solutions to a problem.

therefore from consciousness that assemblages and structures, bodies and materials derive like so many *mechanai*, at once machines and expedients.

Morphogenesis, for Ruyer, is not the activation of a previously given structure but the appearance of structures. This is why it does not coincide with functioning either. Functioning is the setting in motion of a structure that, consequently, must pre-exist it, formation being firmly linked, as in Leibniz, to the unity of a development that, however, unlike Leibniz, Ruyer thinks of as the unity of an undivided rather than an individual or individualised being. This unity, i.e. consciousness itself as the primary bond, is what Ruyer calls "absolute survey", meaning existence together as an immediate form [58, p.107], a unity that is the very set of details survey at a speed that must be assumed infinite, or at any rate very high, given that its non-reproducibility, according to the fathers of the *Extended Mind* [16] is a function of the amplitude of its band and the very high frequency of its communications [17, p. 7][22].

For Ruyer, the absolute domain of survey is the pattern common to the consciousness of the molecule, the atom and the organism insofar as each of them is a multiplicity of non-local unobservable links having a unity irreducible to the *relata* in the form of a being-together [60, p. 107]. Yet, given the non-dimensionality of the survey, the *multiplex unity* of consciousness is not realised by virtue of something that, from the outside, collects the data of the multiplicity in deference to the law of perception: n +1. At a primary level, consciousness is not representational, nor is it, after all, subjective in the strict sense, i.e. egological (something Bergson also suggests when, as we have seen, he specifies that the *élan* is not "the reduced consciousness that dwells within each of us"). As absolute domain primary consciousness is rather, a subjectivity without a subject, because it is not the subject that makes consciousness possible: quite the opposite. Consciousness is an absolute surface insofar as it is one-sided, non-cloneable, without reverse and edge, i.e. without beginning and end. This means that, unlike the *res extensae* that crowd the space of geometry, it is not given to anyone, nor is it included in this space, since it is the origin of this space. The laws of transcendental aesthetics and the mechanics that go with it—a phantom, according to Ruyer—do not apply to it because consciousness is present everywhere as a local absolute of existence rather than presence.

6.5 Conclusions

In the ninth chapter of *Neofinalism*, Ruyer offers an example of primary consciousness by resorting to the analogy with the structure of the visual field: the visual field, he explains, is survey by consciousness in an immediate way in the sense that its details or elements are not linked together by causal connections, like the parts of a machine,

[22] However, given the ubiquity or non-locality of consciousness, the problem is the existence of localised zones rather than the one of "wide" localisations: the latter are the primary datum [58, p. 101].

but are grasped as a whole with an absolute temporal and spatial survey which is absolute in so far as it is independent of any additional dimension (the formula of the field is n = 1). Where, as a rule, analyses of consciousness presuppose the necessity of an external observer, a circumstance that opens the gates of regress to infinity, Ruyer introduces a type of consciousness that arrests regress insofar as it is intuitable without a third dimension. With respect to it, distance is meaningless, which is why the absolute character of the domain—a survey without survey—guarantees that the absolute phenomenality disclosed by the seventh thesis of the second hypothesis *of the Parmenides* does not resolve itself into a whirling *myse en abime*. Indeed, for the same reason that we cannot circumscribe the field of vision by reducing it to one thing seen among others, we cannot see this consciousness otherwise than by being it, and it is the tautohegorical statute of the absolute survey which, like the oriental divinities according to Schelling or Plato's ideas, signifies only itself, that averts the danger of a third man or third eye.

Primary consciousness is at the same time, fact and act, content and law, the poplars and their spinning or, to take up the examples of the essay of Sarti and Bifo, the whips and ecstasy [63, p. 45]. But also, a pure form that is, at the same time, pure apperception. To grasp its being is to merge with it because being a consciousness and having a consciousness, for Ruyer, coincide in the same way as having a feeling of extension means being extended. The absolute surface is not a surface that we survey but a surface that we are, and what Bergson calls the "immediate data of consciousness", in Ruyer, become, because of this, consciousness itself as an immediate data, i.e. something which is not given, properly, to anyone. *Consciousness is, rather than being given or known, and is in itself*. The extension known to someone, or sensible in the Kantian sense, occurs through the spontaneous collapse of consciousness as a true extension, being of the sensible, under observation, every field, including the conscious one, being a kind of multi-spatial environment in which the phenomenon of the "out of place" occurs. Ruyer says that it is seen whenever it is known by another being, when an inspection objectifies or corporatises it. As for *itself*, however, the field is *visio sui*: a mirror that mirrors itself.

What we call the body, for Ruyer, is consciousness as *esse in se* observed externally and created, by this observation, as *esse in alio*: nothing more than the mechanical guise that organic consciousness takes on, viewed from the outside. The soul is the form itself that is observed as the body [56, p. 101]. The *res extensa* is only «the byproduct of the perception of a being by another being [58, p. 77]: an epiphenomenon to the contrary, considering which, Ruyer remorselessly sentences that there is no body [il n'y a pas de corps] (Ibid., p. 77). Matter, in fact, is simply the giving of form to form because the fundamental mode of being, even when matter appears, is and remains existence as an absolute domain: not an extended localised in space and time, but a delocalised and delocalising field of consciousness that, at every instant, is in every point of itself as a background that never becomes a figure. The field is seen in the sense that it is a vision of itself, and a vision is not itself a thing seen.

However, it is precisely the unobservable nature of unitary fields that relentlessly fuels scepticism towards monadologies or their truth: panpsychism. Ruyer combats it by insisting that, whether one considers the molecular, cellular or organic level, coordinating units, albeit invisible, must be admitted if one is to explain the harmony, precision and extraordinary synchronisation with which embryos "manage" to produce a complete organism despite the experimenters' manipulations, ablations and transplants. Not only, in fact, does the egg have the power to form complex organisms: the way it does so gives the distinct impression that it actively "strives" to create it knowing, moreover, what it is doing while it is doing it. Primary consciousness is this organic, unconscious and automatic knowing. The need to identify it with a self-forming form becomes even clearer when embryogenesis is considered. This only makes sense if the embryo has/is this consciousness, because only if it has/is this a "self-forming" unity ("self-enjoyment" is the term most often used by Ruyer to qualify this self-possession or self-fruition[23]) can an embryo evolve in all dimensions of space–time, maintain equilibrium and proportions while changing dimensions, and restore the general pattern of the organism when it is damaged. Its capacity for invention and adaptation, its ability to realise certain patterns specific to its species, as well as its simultaneous development on multiple scales and in multiple directions cannot be explained either by mechanistic materialism or the classical preformism of the late nineteenth century because the egg is not the adult organism in power, and this is not contained in miniature in the first.[24]

In Ruyer's eyes, the facts of equipotentiality, of the multiscale spatial unfolding of developmental processes, together with the paradoxes of pattern formation and cellular differentiation based on divisions of the same original cell, all point to a non-spatial and non-localisable origin of the information that drives morphogenesis. The latter is akin to an informed formation that proceeds recursively by *loops* governed by axiological feedback on a trans-spatial *eidos* that, in Ruyer, takes the place of the demiurge[25]. Creation that is not a mystery, for him, is an unbridled thematicism that leads him to call his monism of subjectivity a philosophy of verticalism [60, p. 11 ss]. His monads are extended as non-spatial spaces generated by the enjoyment of an ideal motif that escapes horizontal relations (the superficial interactions that occur *de proche en proche*, as Ruyer likes to say, i.e. by contiguity and proximity). Its unfolding spreads, rather, a domain that possesses itself from within: an expressive plane that generates and regenerates itself in a self-normative dynamism capable of creating effects of projection—secondary consciousnesses, bodies, materials and

---

[23] Ruyer says he took the term "self-enjoyment" (which he always retains in the English original in his work) from Samuel Alexander's *Space, Time and Deity* (1920) and not from Whitehead. But Whitehead did the same.

[24] For De Landa, in the transition from the egg to the organism there is a continuous cascade of phase transitions with symmetry breaks: a progressive differentiation by which one passes from a non-metric to a metric space through a hierarchy of geometries differing by increasing degree of differentiation: from topological geometry, the least distinct, to Euclidean, the most distinct, via differential, projective, affine (see [23, pp. 11 ff.].

[25] For Sarti and Bifo, too, there is an *eidos* in the sense of an idea generating form, of an algorithm leading to form, of a form forming form. This form governs formation by selecting between the various potentialities without prejudice to the immanent creation of conditions by the morphogenetic process itself: conditions, in fact, belong to becoming form insofar as they are produced by it.

structures—while resisting all projection (the surface is absolute because it is not specular).

To be for oneself extended means in fact that the having-place of enjoyment creates its own place and that this creation is the ex-tension of an intensity in a spatiality before space, and thingness before the thing, in which the artificial formality of both is announced by non-metrical ties that consciousness improvises in accordance with the type that dominates it and exists as a vague but not amorphous directionality, preliminarily, albeit only in a logical sense, with respect to the execution of the thing-extended-in-space structure thus created. Conceived in these terms, morphogenesis is the conquest and creation of space, as well as of time, from which, however, it does not detach itself. Space and time are not *pre-formed* but *performed* by dramatizing, sensifying, those "distinct-obscure" elements [24, p. 240; 281 ff.] that are the generative differentials of non-metric multiplicities studied by Riemann and Deleuze: sets populated by vector fields—linked by bifurcations that break their symmetry—and by distributions of attractors that define their levels. And since dramatization requires a bodily depth of application, given that formation needs residuals of form from which to relaunch itself, morphogenesis is always a creation of forms concomitant with the bending of existing organisational constraints. Or rather: since the creation of forms is directly proportional to the capacity of the elements, which up to that moment were distinct, to lose further individuality through thematic participation in something that belongs to everyone without belonging to anyone, it is the bending, that is the loss of individuality through sharing a theme-type, that is the true creation.

References

1. Auger, P.: L'Homme microscopique: essai de monadologie. Flammarion, Paris (1952)
2. Baumgarten, A.: Aestethika. Meiner, Hamburg (2007) (1750)
3. Bergson, H.: Time and Free Will. Essay on the Immediate Data of Consciousness. Dover, Mineola-New York (2001) (1889)
4. Bergson, H.: Matter and Memory. Zone Books, New York (1988) (1896)
5. Bergson, H.: Cours au Lycée Henri IV. In: Id. Lezioni di metafisica. Spazio, tempo e materia. Mimesis-Milano Udine (2018) (1893)
6. Bergson, H.: Creative Evolution. Macmillan, London (1944) (1907)
7. Bergson, H.: Introduction to Metaphysics. Putnam's Sons, New York (1912)
8. Bergson, H.: Duration and Simultaneity. Bobb Merrill: Indianapolis (1965) (1922)
9. Bergson, H.: 1932. The Two Sources of Morality and Religion. Greenwood Press, Westport-Connecticut-London (1974)
10. Bergson, H.: The Creative Mind. Philosophical Library, New York (1946)
11. Bergson, H.: Histoire de l'idée de temps. Cours au collège de France (1902–1903). Puf, Paris (2016)
12. Berkeley, G.: The analyst. J. Tonson, London (1734)
13. Bitbol, M.: Philosophie quantique: le monde est-il extérieur? Mimesis: Sesto San Giovanni (MI) (2023)
14. Bryant, P.L.: Difference and givenness. Deleuze's Transcendental Empiricism and the Ontology of Immanence. Northwestern University Press, Evanston (2008)
15. Cassirer, E.: Das Erkenntnisproblem in der Philosophie und Wissenschaft der neueren Zeit. Bruno Cassirer, Berlin (1920)

16. Chalmers, D., Clark, A.: The extended mind. Analysis **58**(1), 7–19 (1998)
17. Chalmers, D.: Foreword to Clark, A. Supersizing the Mind: Embodiment, Action and Cognitive Extension. Oxford University Press, New York (2008)
18. Chatelet, G.: Les enjeux du mobile. Mathématique, Physique, Philosophie. Seuil, Paris (1993)
19. Chatelet, G.: To Live and Think Like Pigs. The Incitement of Envy and Boredom in Market Democracy. Urbanomic and Sequence Press (2021)
20. Clagett, M. (ed.): Nicole Oresme and the Medieval Geometry of Gualities and Motions. A Treatise on the Uniformity and Difformity of Intensities Known as 'Tractatus de Configurationibus Qualitatum et Motuum. University of Wisconsin Press, Madison, Milwaukee (1968)
21. Cohen, H.: Das Prinzip der Infinitesimal-Methode und seine Geschichte. Ein Kapitel zur Grundlegung der Erkenntnisskritik. Suhrkamp, Frankfurt/M. (1968) (1883)
22. Colonna, F.: La monadologie de Ruyer. Revue de Métaphysique et de Morale **3**, 321–332 (2020)
23. De Landa, M.: Intensive Science and Virtual Philosophy. Continuum, London (2002)
24. Deleuze, G.: Difference and Repetition. Columbia University Press, New York (1994) (1968)
25. Deleuze, G.: Francis Bacon. Logic of Sensation. Continuum, New York-London (2003) (1981)
26. Deleuze, G., Guattari, F.: Anti-Oedipus: Capitalism and Schizophrenia. University of Minnesota Press: Minneapolis (1977) (1972)
27. Deleuze, G., Guattari, F.: A Thousand Plateaus: Capitalism and Schizophrenia II. University of Minnesota Press, Minneapolis (1987) (1980)
28. Detcheverry, T.: Deleuze Lecteur de Leibniz: la distinction de l'espace et de l'étendue. La Deleuziana **11**, 56–70 (2020)
29. Freud, S.: Beyond the Pleasure Principle. Norton & Company, New York-London (1961) (1920)
30. Freud, S.: Findings, Ideas, Problems. In: The Standard Edition of the Complete Psychological Works of Sigmund Freud, vol. 23, pp. 299–300 (1938)
31. Goff, P.: Consciousness and Fundamental Reality. Oxford University Press, New York (2017)
32. Goff, P.: Galileo's Error: Foundations for a New Science of Consciousness. Pantheon Books, New York (2019)
33. Gueroult, M.: L'espace, le point et le vide chez Leibniz. Revue Philosophique de La France et de l'Étrange **136**(10/12), 429–452 (1946)
34. Gunter, P.A.Y.: Bergson, Mathematics, and creativity. Process Stud. **3–4**(28), 268–288 (1999)
35. Heidegger, M.: The Question Concerning the Thing. On Kant's Doctrine of the Transcendental Principles. Rowman & Littelfield, London-New York (2018) (1962)
36. Henry, M.: The Essence of Manifestation. Martinus Nijhoff, The Hague (1973)
37. Henry, M.: Material Phenomenology. Fordham University Press, New York (2008) (1990)
38. James, W.: Bergson and his critique of intellectualism. In Id. A Pluralistic Universe. Arc Manor, Rockville (2008) (1908)
39. James, W.: Essays in Radical Empiricism. Longmans, New York (1912)
40. James, W.: Some Problems in Philosophy. A Beginning of an Introduction to Philosophy. New York, Longmans (1921)
41. Kant, I.: Critique of Pure Reason. Cambridge University Press, Cambridge (1998) (1781)
42. Kant, I.: Anthropology From a Pragmatic Point of View. Cambridge University Press, New York (2006) (1798)
43. Koyré, A.: La loi de la chute de corps. Galilée et Descartes. Revue phiosophique **23**(5/8), 149–204 (1937)
44. Leibniz, G.W.: New Essays on Human Understanding. Cambridge UP, Cambridge (1966) (1765)
45. Leibniz, G.W.: Philosophical Essays. Hackett: Indianapolis & Cambridge (1989)
46. Lyotard, J.-F.: Anima Minima. In id. Postmodern Fables. University of Minnesota Press: Minneapolis, pp. 235–249 (1997) (1993)
47. Maier, A.: Das Problem der intensiven Größe in der Scholastik: De intensione et remissione fomarum. Heinrich Keller, Leipzig (1939)
48. Maier, A.: Die Mathematik der Formlatituden. In: Id, pp. 257–384. An der Grenze von Scholastik und Naturwissenschaft. Essener Verlagsanstalt, Essen (1943)

49. Maimon, S.: Gesammelte Werke. Olms: Hildesheim (1965)
50. Miller, A.I.: Imagery in Scientific Thought. Creating 20th-century Physics. Birkhauser: Boston (1984)
51. Miravete, S.: Bergson structuraliste. L'aube d'une autre anthropologie structurale et cognitive. Paris, Les Presses du Réel (2023)
52. Peirce, C.S.: Collected Papers of Charles Sanders Peirce, vol. 1–8. Harvard University Press, Cambridge (1931–1958)
53. Philonenko, A.: L'Œuvre de Kant. Vrin: Paris (1969–1972)
54. Philonenko, A.: L'école de Marburg. Cohen, Natorp, Cassirer. Vrin, Paris (1989)
55. Ruyer, R.: Le versant réel du fonctionnement. Revue Philosophique de la France et de l'Étranger **5/6**(119), 338–362 (1935)
56. Ruyer, R.: La conscience et le corps. Alcan, Paris (1937)
57. Ruyer, R.: Parallélisme et spiritualisme grossier. Revue philosophique de la France et de l'étranger **125**(1–2), 110–127 (1938)
58. Ruyer, R.: Neofinalism. University of Minnesota Press, Minneapolis (2016) (1952)
59. Ruyer, R.: Leibniz et M. Tompkins au pays des merveilles. Revue Philosophique de La France et de l'Étranger **147**, 27–40 (1957)
60. Ruyer, R.: The Genesis of Living Forms. Rowman & Littlefield, London-New York (1957) (2019)
61. Ruyer, R.: Paradoxes de la conscience et limites de l'automatisme. Albin Michel, Paris (1966)
62. Ruyer, R.: L'embryogenèse du monde et le Dieu silencieux. Klincksieck, Paris (2013)
63. Sarti, A., Bifo, F.B.: Run. Forma, vita, ricombinazione. Mimesis, Milano-Udine (1997)
64. Sarti, A., Citti, G.: A fistful of differencials. La Deleuziana **11**, 4–18 (2020)
65. Sarti, A., Citti, G., Piotrowski, D.: Differential Heterogenesis. Mutant Forms, Sensitive Bodies. Springer, Berlin (2022)
66. Simondon G.: Individuation in the Light of the Notions of Form and Information. University of Minnesota Press, Minnesota (2020) (2005)
67. Sylla, D.: Medieval concepts of the latitude of forms: the oxford calculators. Archives d'histories doctrinale et littéraire du Moyen Age **48**, 223–283 (1973)
68. Valery, P.: Discours sur l'esthétique. In Id. Varieté IV. Gallimard, Paris, pp. 235–265 (1937)
69. Vermeiren, F.: The Leibnizian lineage of Deleuze's theory of the spatium. Deleuze Guattari Stud.**15**(3), 321–342 (2021)
70. Viveiros De Castro, E.: La forêt des miroirs. Quelques notes sur l'ontologie des esprits amazoniens. In: Laugrand, B.F., Oosten, J.G. (ed.), La nature des esprits dans les cosmologies autochtones. Presse de l'Université de Laval, Laval, pp. 45–74 (2004)
71. Whitehead, A.N.: The concept of nature. Cambridge University Press, Cambridge MA (1920)
72. Whitehead, A.N.: Process and Reality. An Essay in Cosmology. The Free Press, New York (1978) (1929)

Chapter 7
MODUS

Riccardo Panattoni

Abstract The essay explores an ontological rethinking of the living as modulation rather than as form or substance. Drawing from Simondon and Deleuze, it proposes a shift from the logic of being to that of modes—dynamic, metastable configurations of existence. Life is not a thing, but a rhythmic variation of intensities that traverses and transforms both bodies and environments. The subject emerges not as a fixed identity but as a singular rhythm, a temporary crystallization within a field of differential relations. Modality becomes the key to understanding processes of individuation, perception, and ethical becoming. Against essentialism and structuralism alike, the living is thought of as a diagrammatic field, where sense, force, and relation are co-constitutive. Thought itself is redefined as an activity of navigation and resonance, always exposed to contingency and relational drift. *Modus* thus calls for an ontology capable of honoring transformation without totalization.

7.1 Preface

Through the concept of *mode*, or better still through the instance of a *modal ontology*, we will attempt—taking as primary references the works of Gilbert Simondon and Gilles Deleuze—to highlight a series of paradigmatic shifts, in order to outline the necessary passage from the concept of structure to that of *field*. The aim is to propose "batteries" of concatenations as points of reference to go beyond the classical philosophical opposition between "being" and "becoming", and move toward the institution of a *becoming of being* as a *dynamic tension in a point of arrest* or as *the opening of a counter-effectuation on the continuous point of emergence*. It is the emergence of a disjunctive synthesis between spontaneous genesis and creative act. The battery of semantic concatenations, in addition to the aforementioned *mode* and *field*, will be composed of the following entries: *singular*, *impersonal*, *virtual*, *counter-effectuation*, *potency*, *diagram*, *fold*, *pre-individual*, *transindividual*, *metastability*, and *transductive*.

R. Panattoni (✉)
Università di Verona, Verona, Italy
e-mail: riccardo.panattoni@univr.it

S. Vizzardelli and R. Panattoni (eds.), *Plasticity and Discontinuity*, Lecture Notes in Morphogenesis, https://doi.org/10.1007/978-3-032-03119-8_7

As an emblematic foundational reference, we could say that if the concept of archive is based on the stratification of knowledge, that of diagram centers on its remaining potential, in a semi-aleatory or semi-random virtuality. Every diagram is thus strategic, while every archive is stratified; every diagram is microphysical or differential, while every archive is macroscopic. The power relations between strategies and strata nonetheless imply form, but concern unformed matter, non-formalized functions. For this reason, at the level of microphysics, it is necessary to abandon the word "structure". From a structuring process, with all its corresponding stable forms, or all the forms that are eventually possible, we move to a permanently dynamic *field of forces*, to a *dynamism of links or knots*. This is about going beyond the strata in order to traverse the unstratified matter. The Outside of the strata, which constitutes them, is the *diagram* that never ceases to mix the power relations that *virtually* outline its field. For this reason, not only are human beings strategies, but also particles, electrons, are strategies of force *fields*. What defines the momentary dynamism of a *diagram* is therefore the potential strategy of a field of forces, whether human or not. Exiting stratifications, traversing them, means reaching the properly oceanic, non-terrestrial element, the unstratified element, the strategic element, insofar as every strategy is oceanic and every oceanic element is a *void* of propulsive potency. On the other hand, as Deleuze reminds us, there are no archives of the sea, there are only port archives, as the sea is pure potency. Water is in fact molecular, that is, empty, while lacking nothing. For this reason, variations cannot come from form, but only from space and time: the various figures of the *diagram* are the *mode* of abstract, *virtual* action within an abstract, *virtual* space–time. In the *diagram*, consequently, there are always more figures, more power relations, stretched within *transductive* links and in a permanent balance of *metastability*. This is why the *diagram* does not correspond to a formation, but to a state of continuous transformation: it is always intermediary, unstable, or more precisely *metastable*. It is also unstratified, but *interstratic*. The *diagram* is also informal, proceeding in *transductive* terms, from point to point, in a multipoint *field* of forces. We could then indicate the *mode* of becoming of the *field of forces* through the neologism coined by the poet Andrea Zanzotto: "*biologale*" [36].

7.2 Ontogenesis or Individuation

In the introduction to his work on individuation, Gilbert Simondon [25, 27] immediately warns that granting ontological privilege to the constituted individual means mistakenly believing that one is engaging in true ontogenesis. This is because the very notion of a principle seems to imply the presence of a characteristic capable of prefiguring individuality as it will be constituted, thereby projecting onto it the properties it will eventually possess once it is actualized. To seek the principle of individuation in a sphere that precedes its actual being means to consider individuation in terms akin to what is conventionally understood as a proper ontogenesis. For this reason, in order to outline its perspective, it becomes necessary to imagine that it is possible to position oneself somewhere "before" form was acquired.

In fact, it means identifying such a principle in both matter and form, and consequently presupposing the existence of a linear succession of time. In other words, it implies asserting with certainty the existence of an origin in which the principle of individuation resides—an origin from which this principle will only later be effectuated within the operation of individuation. Yet in reality, every individuation—remaining within the principle of its own becoming—is always already *informed* by its own mode of being. This is why the concept of "principle" cannot be conceived simply as synonymous with the concept of origin, as an initial moment capable of identifying in its constituted realization the destiny toward which it was always already directed: an initial moment that already contains within itself what the individual will eventually manifest as such, thereby accounting for its very haecceity. On the contrary, the term "principle" must be understood as that which sustains in act, in the actuality of its unfolding, a continuous movement of *counter-effectuation*, such that its initiality continues to pulse in time as a permanent form of potential emergence, oversaturated by its own *mode* of being.

Consequently, there is always a blind spot, a generative space, that conceals every constitutive operation of individuation. A blind spot which, in the case of the human process of subjectivation, we might refer to as the act of the unconscious. This is why Simondon insists on the concept of the *pre-individual*, which must not be understood as what precedes the individual, but as what cuts back against the continuous movement through which individuation tends to unfold, effectuating it as a becoming of being.

On the other hand, if we consider that states are *modes* of *metastable* being, then being and becoming no longer appear as opposed notions, as abrupt transitions from one structure to another. The concept of becoming is no longer the expression of the continuity of a process of alteration, but rather a concatenation of tensive states through the release of potential energy, whose existence forms part of the causal regime that constitutes the *virtual* transition of momentary states even as they vibrate within themselves. In such a way, the energy contained in the system of *metastability*—through which individuality continually takes form—is the same that actualizes itself in the transition from one state to another: this is what Simondon defines as a "structure-energy", and that we, following Deleuze [7] might describe as a *diagram*. Thus, the unity of structure is always more than one, without necessarily becoming two. It is a tensive *modularity* that goes beyond the need to uphold the actual unity of the one, which does not mean, however, that it is also necessary to deny the distinctive trait of homogeneity of the process of individuation. This *modal ontology* makes it possible to avoid falling into the classic distinction between the transcendental and the empirical, and, with regard to the human being, it prevents us from recognizing it solely through the attribution of an essence. The living being, in fact, never presents itself as a fully individuated whole, because its individuation remains a *diagrammatic* passage that preliminarily cuts through all its features, keeping it open to the need of a *pre-individual* passage of being: its becoming.

The concept of ontogenesis must therefore not simply be expunged from philosophical analysis, it must also be understood as a *metastable*, *diagrammatic* process, which always has priority over the actuality in which the individuated being unfolds over time. Also, classification based on genre collapses in this *mode*, since there is no longer a direct correspondence linking it to its genesis, but rather the unfolding of a transformation of punctual acts that remain alive in their *potency* once their genesis has been effectuated, without the generative act thereby being reduced to a mere "*incipit*", an initial push for what might then be considered a continuous and linear process. The *pre-individual*, *diagrammatic* reality, in relation to the concept of genesis, thus never ceases to re-traverse, to cut through, the individuated being, in such a *mode* that its *potentiality* cannot be resolved or fixed within an exclusive structural system. The process of punctualization, in which the individual is actualized, therefore does not correspond to what might be understood as a constituted individual, but always also reveals a slippage, a lateralization. So much so that the *pre-individual* state of being continues to subsist alongside the result of each single point of individuation, in a *mode* by which the living being itself retains within itself a punctual activity of permanent individuation, giving rise to the simultaneity of what we might describe as a true theater, a radial array, of individuations. It is a regime of concatenations that take on the dimension of an "internal resonance" which, through its constant act of "informing", maintains the very condition of the living in a state of *metastability*. The latter, in fact, does not resolve problems merely by adapting, that is, by modifying its relation to the environment, but by also modifying itself at every step, constantly creating new internal structures, seeking *modes* to re-traverse its vital dilemmas. For this reason, Simondon finally asserts that the living individual constitutes a system of individuation, an individuating system, and a system that individuates itself, which function in the simultaneity of their disjunctive synthesis [4]. The becoming of the living is consequently a *modulation* that takes place at the limit not so much between the domain of interiority and exteriority, but rather between the Inside and the Outside.

7.3 The Outside and the Inside

It is not possible to determine a process of individuation by extracting it from the field of forces from which it is constituted and which it, in turn, helps to constitute: every individuation is an expression of the energy through which the field of forces does not so much structure it as establish it as a *fold*. However, insofar as we think of this determination in terms of forms, it becomes necessary to refer to the dimension of "knowledge" and consequently to the related concept of exteriority, more than to the concept of the Outside. The latter comes into play when one shifts from taking forms into account to confronting the immanent dimension of forces—which is when the concept of exteriority no longer applies, and the concept of the Outside emerges in all its formless character. For the Outside is expressly the formless index of forces, the oceanic whole of potential intensities. It corresponds solely to the fluid—*virtual*—totality of a state permanently in motion: an ocean of data and information that moves while remaining in its own state at a strictly molecular speed. This *potency* of the oceanic zone is anything but foreign to the singularity of the living, for it constitutes its Inside, inasmuch as the living being determines at every point of its individuation process a being that is slow, suspensive, in relation to the speed of the Outside. In this sense, the movement of the Outside causes the living being to fold back onto its own temporality, and once folded, to constitute not so much the concept of an "interior"—still typical of forms, which would require a corresponding knowledge—but the Inside of the Outside, which makes of this *fold* the power of a "non-relation" of the individual with itself: its act of the unconscious. In such *mode*, the *fold* of the Outside thus comes to outline a "zone of subjectivation" that will not have been the self of an "I", but the wave, the ripple, the punctually momentary trace of what is alive: a living present.

We could then say that the *fold* is a momentary stratification in which the Outside finds its surface, a rhythmic movement that brings life to the particular edge of its own determination, and the wave is the *fold* in act, in which the potency of the oceanic zone makes itself felt, touching that singularity in its Inside, and rendering it the breath of a "living present". There is, consequently, no oceanic zone without the singularity of that momentary wave—or rather, to refer to the oceanic zone beyond that particular point which the singularity of the wave determines is to point to an abstract or merely conceptual totality. In other words, it means creating a concept capable of remaining permanently available and repeatable in support of a metaphorical idealization between form and matter. It is in this sense that Deleuze and Guattari, following the work of Franz Kafka [6], propose to abandon the concept of metaphor in favor of that of metamorphosis. So although the concept of Inside remains conditioned by the concept of Outside, which has priority, the Outside, in turn, can subsist in the *virtuality* of its potency only thanks to the momentary discontinuity of the Inside, to the living impersonality of its singular breath, constantly immersed in its immanent principle [10].

After all, if we consider the human being—as a pivotal point that most directly engages our particular attention—singularity does not arise from the child's exiting the mother's body, but from the cutting of the umbilical cord. The child is born into its singularity by bringing Inside itself the air of what is Outside. Later, over time, it will also learn to walk, to move through linear space in order to eventually reach its goal, its purpose; but along that path, also when it will have grasped that object with its hands, but also when it will have not, it will never have stopped breathing, it will never have ceased being traversed by one breath after another. Each single breath is thus the point where the distinction between continuous and discontinuous collapses, between active and passive, between voluntary and mechanical. It is similar to the intensive and dynamic movement of diastole and systole, where the heart contracts and pulses, contracts and pulses, like the rhythm of life: between singularity and impersonality. Each time, it is a different breath, but ultimately it is always also the same breath. The umbilical knot that ties our body to itself never stops reminding us of this. Even though our body moves with confidence in its upright posture, it never ceases, through that knot, to rotate simultaneously around itself. An *umbilicus* of the body that is our *mode* of remaining linked to the non-appropriable memory of the initial breath, as if each new breath were reaching out in an immemorial search for that first moment—or at least toward its phantom. That moment cannot in fact be defined as a truly initial moment; it does not at all correspond to what we tend to define as a point of origin. Since that body was already breathing Inside itself with the breath of the mother's Outside, and that wound, the cut of one's body upon the body of the mother, existed before the child, before the mother, without for that reason being able to withdraw from the force of their inescapable and mutual singularity.

What we are generally inclined to indicate as what is internal is in reality always the Inside of an Outside; it is never what could be understood as a form of interiority. On the other hand, only an Outside can have an Inside, a folding, a *fold* that is the breath of an invagination of life. The moment the line of the Outside *folds*, it establishes an Inside that is the motion of waiting or of exception. And the different *modes* of folding will be the theatrical batteries of the different *modes* in which subjectivation will unfold, since subjectivation never responds to a principle of uniqueness; its Inside is deeper than any form of interiority: it is the *fold* of an unthought of thought, the mimetic double of a *counter-effectuation* of the subject upon itself. This is why, between visibility and statements, despite continuous intersections, the process of a disjunction is always at work. The formalization of words and images, the knowledge that underlies them, is continually traversed by the oceanic forces of visibility and vocal tonalities, in a *mode* that makes visibilities correspond to the call of words, to their statements, even if what is being spoken of is never actually seen, while words correspond to the suggestion of visibilities, even if what is actually seen is never truly spoken of. Their continuous disjunctions therefore do not correspond to a form of separation, but to the demand for continuous virtuous knots; they respond to the *virtuality* of what we might call *diagrammatic folds*. So that if a strategy is given, this strategy is oceanic; the *folds* are oceanic cartographies that never draw upon archival knowledge.

The Outside is therefore the field of forces to which power responds. The latter, however, must be understood in its double meaning: as that which tends to impose itself by force, and simultaneously as that which remains in a *potential* state. Forces push their actualizations toward power, even if this is, in certain respects, a blind push, lacking any intentionality, since in the oceanic zone of life, forces do not all tend in the same direction of what is actual, but rather contrast each other, creating a *field* of permanent *metastable* equilibrium. It is therefore not possible to think of the *field* of forces as a classical transition from potential to actual [1], since for the living being everything is already actual, with no temporal space in which to establish a point of origin or a pure beginning: life remains immersed in the exclusivity of its full immanence. For this reason, the forces correspond to the power of that which tends "in itself" to affirm itself, even though, at the same time, the oceanic zone also keeps in play the forces in their remaining *potential*, in the momentariness of an intensification that never fully resolves itself in its actualization, but keeps alive "in itself" the virtuality of what insists on its "step", on its "yet again" [22].

It is in the dispute between forces within their *field of immanence* that the undecidable contrast between continuity and discontinuity unfolds—a contrast always in becoming between potential and actuality, which drives the human living being to give form to the testimony of its own presence, its own being-there, constantly attempting to reduce the Outside to an exteriority and the Inside to an interiority. Forms are in fact the attempt, on the part of the living, to surpass the indistinct forces of the oceanic zone in which life incessantly sinks into itself. And it is precisely by acquiring form that the human being outlines the *diagrams* of its own knowledge, that is, it seeks to codify its technical ability and to transpose this ability into the creation of innovative technological objects [26, 29].

7.4 The Absolute Outside or the Non-Relation

Diagrammatic knowledge thus responds to the attempt to give form to the constitutive link between the principle of reality and life, in the attempt to bring such form to a permanent stability, even though the Outside remains, nonetheless, a relation of forces in the oceanic zone of life, where forces meet or clash in an intensity whose balance knows no other structure than that of a propulsive *metastability*, which cannot be formalized. The relation of force with life thus gives rise to a diagrammatic force that takes form, folds, becomes an oceanic wave. Every *diagram* is thus always a form traced upon the propulsive force of the Outside. In these terms, however, the *diagram* maintains a relative tension with respect to the relation of forces of the Outside—an inevitable relativity with regard to the necessary involvement of the living being, especially the human living being. This does not mean that the living being cannot experience an encounter with an absolute Outside or an immediate Outside. It does encounter it, however, in the form of a "non-relation", since the "non-relation" does not belong to the order of the negation of relation, but rather to a form of absolutization of relation itself. The "non-relation" with the Outside is the loss of every diagrammatic form, since it continually exceeds the positive by destroying it, as one is already involved in it through an experience that initiates more than any possible initiative, one that anticipates every beginning and excludes any action capable of emancipating itself from it definitively. Deleuze defines this relation without relation with the Outside as the passion that does not allow itself to be mastered in patience—it is the passion of the living being seized by the Outside as such.

The absolute line of the Outside, where passion traverses the living being, is the line of "one dies". This articulation of the absolute Outside as "one dies" is taken up by Deleuze in dialogue with Maurice Blanchot [2] who emphasizes how death is fundamentally twofold: there is a death that is stated in the form of "I die" and a death that is stated in the form of "one dies". The "I die" is death as an indivisible moment, it is the death that can happen to us, that will happen to us. It is a certain fact, caught in the order of temporal possibility, but inescapable from a personal point of view. The certainty of this fact, however, collides with the equally certain fact that one is alive. This is why the temporal structure keeps "I die" within the order of the possible, while "I am alive" remains within the temporal order of a fully immanent experience. Meanwhile, the death of "one dies" remains a dying coextensive with life, a death that has already always begun and that never ends. It is the continuum of a becoming-event that concerns us, that we encounter through the intensity of discontinuous singularities, but with which we are unable to establish any possible relation. It is the non-relation with death in an encounter, that cannot be grasped, with "one dies", or rather, we could say that we are able to encounter "one dies" only because life "is lived". A death, a disappearance, is never one's own death, because the event remains non-appropriable: it is an abyss of the present, a time that seems to be without a present and with which no possible relation can be established—only an absolute relation. A non-relation toward which we are, however, unable to abandon ourselves, because in such an encounter we do not die. In this death of "one dies", singularity is woven into the continuous impersonal procession of the "one". This line of the Outside is something even deeper than "one sees", "one speaks", or "one fights". It is a "one" to which it is not possible to oppose "one lives", because in this continuous stretch of impersonality, "one dies" and "one lives" remain coextensive, though one might say that a minimal trace of discontinuity is "one breathes"—a minimal singularity that nonetheless becomes a *fold*, a trait, which cannot be grasped, of unexpected momentariness.

On the other hand, "one breathes" remains an interstratic phenomenon, occurring within the oceanic movement of "one dies" as a form of "one lives", destabilizing the assumption of an all-encompassing unity of that movement. It is a *folding* movement that stays internal to its own level of reality, without introducing any necessary transition toward something else, yet at the same time keeping its transitive *potency* active: its definitively *virtual* inflection [3, 4, 13, 16, 21]. It is a *folding* that prevents one from grasping processes only in trans-stratic terms, because although it belongs to the passage from one stratum to another, it also always remains partially localized, relative to that particular domain of singularization. The interstratic, *transductive* trait remains, each time, in its consistency as an event, a momentary passage, a molecular process suspended in relation to any molar, trans-stratic dimension [7], and it lets new, semi-randomly imprinted forms of life emerge. For this reason, Deleuze claims that the line of the Outside must be traversed by a *folding* movement—by its *mode*—because the multiple is not only what has many parts, but also what is folded in many *modes* [8]. It is therefore necessary that the Outside fold into the invagination of an Inside that is more intimate than any form or domain of interiority: an absolute proximity. It is, indeed, the breath of the *fold* of the line between Outside and Inside that tears, in moments of singularization, the coextensive line of "one lives" across "one dies", and situates a life on the suspensive stretch of a transcendental field of pure immanence [10].

We therefore no longer have a transcendental that makes experience possible by linking it to the cognitive structures of the subject [19], but a *pre-individual* condition of experience as affectivity cut into the event. An impersonal *folding diagram* that traverses, in permanent discontinuity, the different intensity fields of what happens: differences and singularizations that exist before any presupposed determination of the subject. These movements, momentary and concatenated *diagrams* between *folds* that turn singular experience of life into a breath that does not belong to the subject, but responds to the non-appropriability of its being a body. On the other hand, however close two sensations may be, the passage from one to the other, the increase or decrease in their remaining in *potency*, their *virtual* quality, does not place immanence in relation "to" substance, but places substance within the *modes* of immanence [31]. In this sense, for life to happen upon itself, to *counter-effectuate* itself in the singularity of a body, in the effectivity of its temporalization [15, 29], in the being of its becoming, the line of the Outside must inevitably become a momentary *fold* of an Inside, rippling the only coextensive line between "one dies" and "one lives". A *diagrammatic fold*, a void of breath, through which all words, all things, all events will also fold in turn: an oceanic wave in the ocean of forces, a momentary ripple that, returning upon itself, will continue to flow, in *potency*, on the surface of depths that are entirely unexplored.

7.5 Counter-Effectuation of the Inside upon the Outside

We could claim that the living individual is contemporary with itself in all its elements, something that, by contrast, is not true of the physical individual, which entails a past that is radically past, even when still in the process of development. The living thus presents itself as a node of informational communication, a system within a system that determines "within itself" a mediation between two different orders of magnitude, insofar as the event that rebounds as *counter-effectuation* onto the living being is nevertheless inseparable from existing conditions, from the bodies and lived experience in which its actualization has allowed itself to be effectuated. What in fact happens, what becomes, always realizes itself in a given condition, but at the same time is *counter-effectuated* each time on the singularity of the living being, abstracting itself from that given fact and liberating the repeatability of the concept. For this reason, Deleuze [5, 23, 24, 30] insists on viewing the philosophical act as an inter-time, as that which sidesteps (like a mime that incessantly doubles the event within itself) the existing conditions through perpetual allusion, but without the need—following Mallarmé—to go so far as to break the ice. Indeed, the mimetic act is not simply limited to reproducing existing conditions, it does not merely imitate lived experience, but creates its singular conceptual image, that which follows from the unexpected development of its *diagram*. If, consequently, thought, in its very unfolding, is called to face an Outside further away than any external world, then what remains closest to it, what thought discovers as its constitutive *mode* of thinking, the *fold* of its Inside, can only appear as the unthought of thought. Deleuze [11] grasps the full scope of this concept by turning to the articulation proposed by Michel Foucault [17] and understanding it not so much as a logical impossibility (as he believes is the case with Heidegger), nor as a vital power (as would instead be the case in Artaud), but as a sort of hallucinatory impotence. This is why the act of *counter-effectuation* does not respond to what could have happened but did not; it does not belong to the order of the possible, but remains as the visual imprint, the becoming-image, of what actually did happen. A return of the living being upon the process of its own subjectivation, which prevents it from fully adhering to the sole continuity in which it finds itself immersed each time. Granting the truth of the event the unique chance not to be confused with its mere realization, and allowing the discontinuity of that living fracture to liberate the event in the hallucinatory impotence of a suspension in any time [8]: in a time that "still" remains in the time that is "yet" to come.

It is in this disjunctive synthesis, in the *fold* between effectiveness and *counter-effectuation*, between actualization and hallucinatory impotence, that the creative act finds its expressive force, the emergence of a *diagram* that, in its becoming a *fold*, is able to flow from point to point: in a *transductive*, diffuse, unstable, and constantly mutating *mode*. The *transductive mode* consists, in Simondon [27, 35], in a state of *pre-individual* tension, that is, in a being that is more than unity and more than identity and that, by continuously relating to itself in *counter-effectual* discontinuity, is not yet articulated in multiple dimensions, in dimensions to come, but remains in the *diagrammatic* emergence of its immanent fact. For this reason, the terms of the *transductive* operation do not pre-exist the *modes* of their own effectuation, but become temporally in a heterogeneity that structures and desynchronizes itself: a discontinuity that articulates the continuity of time in which the fact unfolds and endures in the singularity of a breath. It is a form of abstraction, an exit into itself of the creative act, which is never determined, in its thinkability, in a general form, as if responding to a universal principle, but always varies from singularity to singularity, without, however, ever perfectly coinciding with the exclusive effectiveness in which that singular life would seem to be articulated. In this sense, the hallucinatory impotence of *counter-effectuation* remains purely *virtual*. Because the *folding diagram* that comes into being is always composed of micro-emergences, of continual absences, while it mixes at every instant with what has been and disappears in favor of another relation of forces in becoming. The *virtual* of the *counter-effectuation* remains, in fact, in its permanent non-actuality, within the order of the real: it is the line of the Outside that *folds* into the Inside of a singularity in which the unthought of thought is introduced as the drive toward the creativity of the fact we do not "yet" think. This does not resonate as the idea that the moment might arrive when we will finally think it, because the hallucinatory *virtuality* of what happens, its state of impotence, its *mode* of singularly being articulated, is already present in the very act of thinking. So much so that the "already there" and the "not yet" appear as two coextensive *modes* of temporality that characterize the *fold* of the "living present" in which every singularity continues to take place. Consequently, one does not "yet" think, because one thinks the already here: its fact, its hallucination.

If folding is creating a *virtual* double of what happens, one can say that the *fold* is the point of emergence of a subjectivation as the double of the Outside—keeping in mind that the double is not a simple reproduction of the same, but a repetition of the different. Because the double is not an emanation of the self, but a placing into immanence of a not-myself (Deleuze 2006), the *counter-effectuation* of temporal deviation. For this reason, the double, the counterfigure, is always temporally the *fold* of the Outside, the fold of the waves along the line of the ocean. In such a *mode* that what folds is the temporally "indifferent" line of unfolding. Since in the human living being the *potency* of *counter-effectuation* is automatic and simultaneous with the act of what is effectuated, no opposition is determined between folding and unfolding, but only different *modes* of folding. This means that the forces of the Outside, the deep currents of the ocean, continuously produce affections on other forces, determining the status of unfolding forces, but simultaneously, in the continuity of the clash of forces, each individual force also potentially produces affections upon itself. There is not only pure temporal unfolding, because the singularity of that force does not merely undergo affections from another force nor produce affections on another force, but remains affected by itself, in other words it finds a *mode* of folding onto itself, becoming the discontinuous time of subjectivation, without subject and without object, but only in an immanent relation of forces. It is like a form of visual disparateness, where even though one eye does not coincide with what the other sees, the visual cortex still manages to integrate the two, and the sensation is that of receiving a stereoscopic image that guarantees a sense of depth, while in reality the vision is always double. The fusion mechanism, through complex neurological processes, causes the eyes to converge on external points of interest, allowing the visual cortex to integrate the images from both eyes into one, although in truth it is a recomposed diplopia, which always underlies what we believe is a unitary process: a sort of optical unconscious. The power to undergo affections is like the matter of force, while the power to affect is like the function of force, a function that is however non-formalized and grasped independently of the concrete forms in which it is embodied, the ends it serves, and the means it employs: the physics of action—says Deleuze—is a physics of abstract action. The *folding diagram*, consequently, regardless of the formed substances, the beings and the qualified objects it will penetrate, can give rise to multiple *modes* of *virtual* distribution of its affective *potency*. But what characterizes it is that the *fold* of subjectivation of which it becomes the point of emergence—a ripple on the oceanic expanse of data—also gives rise to a form of waiting, to a temporal deviation from the actual events that occur. It is as if subjectivation, caught in the forms of its becoming *fold*, also kept waiting for itself: a slow being immersed in infinitely high speeds that traverse it.

7.6 Diagrammatic Folds

It is therefore possible to state that between organic and inorganic matter a difference of vector seems to emerge, wherein the inorganic seems to move towards increasingly large masses, where statistical mechanisms operate, while the organic seems to move towards increasingly polarized masses [11], where mechanisms of individuation are virtualized. For this reason, the *virtual* inflection, the singular punctuation that ripples the impersonal multiplicity of data, is what, with Paul Klee [20], we might indicate as the place of cosmogenesis. A non-dimensional point of persistence between the dimensions that are actualized, an event "always" waiting to become once again the event that it is. In such a *mode* that the being-in-itself of *potency* persists temporally in its becoming act and the being-in-itself of the act is already mechanically *counter-effectuated* in the *virtuality* of its becoming *fold*. What occurs is the *diagrammatic fold* persisting in the relationship between its becoming patience and its becoming form, structurally traversing a lack, a becoming void of time within time, despite the continuous unfolding of the linear forces of the Outside.

That is why the oceanic zone of the latter can tend to remain repressed, insofar as individuality may find itself engaged in the exclusive attempt to grasp the unity of itself in time, through its own capacity to create forms, to structure the corresponding knowledge. If individuality thus assumes it is a constituted entity, it will tend to project the "one lives" in a coextensive form, and the *diagrammatic fold*—instead of *transductively* passing from *fold* to *fold*—may aspire to articulate its own permanent design. A design capable of projecting itself so as to overlap with the repression of the absolute Outside of "one dies", contrasting it with the continuity of a creative, vitalistic force that adheres, in a similar absolute way, to the natural principle of life. Whereas in reality every *diagram*, coming from the Outside, does not allow for the identification of a temporal possibility of verifying the existence of a first *diagram*. Because every relation of forces always refers back to a previous *diagram*. There is therefore necessarily an element of randomness in every diagrammatic unfolding, which remains linked to the *pre-individual* character of the living, which has marked it in its discontinuous punctuality. It would therefore be more accurate to say that the *diagram* has traits of semi-randomness, that is, it maintains successive transitions but only partially dependent ones, while also being a source of imperceptible, informal creative discontinuities: the singular, immanent breath of a life. It is indeed true that in his conversations—as Deleuze [12] himself recalls—Francis Bacon [34] states that there are only two risks in painting: illustration and narration. Whereas a painting is first and foremost a field of forces, a field of tension between light and color, which cannot be reduced to its mere representation. If anything, what it represents is that which, in certain forms, was already present on the blank canvas before the painter began to paint, and against which, in order to access the singularity of an impersonal creative act, he had to struggle so as not to surrender the entirety of the field to it. That is, he had to struggle with his own intention to paint something. Because the intentional act, in painting, is always figurative, it inevitably takes on narrative aspects. This is why one should never actually realize what one intends to produce, since the creative act, in order to be such, must experience the catastrophe of intentionality [32, 33]. This catastrophe should not be equated with failure—though that possibility is always present—but rather as the irruption of what Deleuze calls a chaos-germ: the contrasting site of forces from which a diagrammatic force may emerge. And if the *diagram* proves truly operative, if it is entirely singular and impersonal at the same time, if it remains wholly immanent to the field of forces in which its *potency* is immersed, then the pictorial fact may emerge. That is, the *incipit* of a form to come may emerge—a point of origin without a true origin, entirely engaged in the perspective capacity to render visible the blind invisibility of a gaze. For this reason, every *diagram* can only be deeply unstable or flowing; it can only respond to the intensity of a *metastable* equilibrium that persists *potentially* within its own transcendental field [7], where mater and functions will continue to mix in order to constitute emerging transformations. Giving rise to a new model of truth that will no longer be subject to history, nor will it feel capable of mastering it, because it will have completely abandoned any reference to the principle of originality. Rather, it will be about the emergence of a number of creative points, of unexpected conjunctions, of improbable continuities—yet capable of destroying the reality of every previous signification, of evading the pure intentionality aimed at reproducing a pre-existing idea or reality.

It is precisely in this myriad of unexpected points that, for Deleuze, the difference between *diagram* and structure, rendered visible by the diagram, manifests itself. Even though it creates each time a transversal network perpendicular to the verticality of a structural form, nevertheless the *diagram* simultaneously outlines strategies that are very different from mere combination. A *mode* that gives rise to the simultaneous superimposition of a map of density and a map of intensity—a *transductive metastability* that proceeds through primary, non-localizable connections: an immanent, non-unifying cause, a "disjunctive synthesis" that endlessly bifurcates the very fabrics of the concatenations it produces. The *diagram* thus becomes the instance through which the creative act undoes the resemblance of what its intentionality would like to reproduce, allowing the singularity of an image-presence to emerge autonomously. Between the visibility of the gaze and the invisible blindness of the visual act, between the before and the after of the pictorial event, there persists the intensive, *transductive potency* of the diagrammatic force. A line that, continually folding back on itself, splits, *counter-effectuates*, and never closes. That is, it never delineates what we could call an inside or an outside, because it is always the simultaneous *fold* of an Inside that breathes the non-appropriable force of an Outside: like life, like one's own body, like the creative act. This line, having originated before any point on the left edge and continuing beyond every point on the right edge, is without beginning and without end; it has never belonged to anyone, even though it pertains in an unequivocal *mode* to the determination of the pictorial fact. As in the painting of Jackson Pollock—a painting that is anything but abstract—where the line never ceases to be the line that becomes, even as it immediately coincides with the surface it completes. Indeed, in the surface and the line the *potency* of two diagrammatic dimensions is always in act, without the union that characterizes them ever breaking, since it is the intensity of the line that *transductively* surpasses its own *potency*, while continuing to become the insistence of its own *diagrammatic fold*. From the outset this line appears adapted to the surface it creates, aligned with the vertical structure of the painting, yet without the painting, which recognizes that line as its own, ever defining its contour. A becoming-*potency* that, in the pictorial fact, remains temporally without end and prevents the very surface it creates from being resolved solely in its transition to act.

It is at this point that the hand will find it is animated by a blind will, capable of imposing itself on an eye struck by the *fold* of a line it can no longer master—one to which it can impose no rule on, nor recognize any kind of knowledge that presupposes a law. A line which, in becoming a shockwave-like surface, changes direction at every moment, preventing the eye following it from remaining engaged in its constitutive capacity for reading, in what it recognizes and feels able to name:

a true temporal catastrophe of the eye. Better then to close one's eyes immediately, if only for a moment, to accept that it is only in the *potency* of that *folding* act, in its *mode*, that seeing will become possible: "close your eyes and you will see" [14, 18]. A *diagrammatic fold* like a ray of time passing through metastable fields charged with continuous tensions. This is why the *diagram* is always also a modulator, because in order to see, the form of a necessary resemblance will not be reintroduced.

It is rather a matter of the combinatory interplay between the modulation of analog language and the articulation of digital language. Whereas in the digital domain meaning is produced through discrete units that can be combined according to rules of articulation, units that are linked in an orderly manner, as happens, for example, with computer languages, with analog language, understood as modulation, the result is a continuous variation, a fluid transformation. This *mode* means form always emerges from modulations of intensity, as with musical tone or the visul of images. A combinatory process prevents a clear-cut distinction between the continuous and the discrete, directly attributable to modulation or articulation. Painting is, in fact, the modulation of light and color in relation to the articulation through which the motif unfolds, finding its own flat surface: whether in Pollock's line with no figure, Kandinsky's abstract figure, or the figural figure of Cézanne or Van Gogh. At the end of modulation, in the *incipit* of the pictorial fact, what we may call the articulation of a resemblance can be found, a resemblance, however, that has taken shape though means of non-resemblance, by means of the sole modulation of light and color. Only in this way does the *diagram* become a matrix of modulation, while the code becomes simultaneously, each time, a single matrix of articulation.

References

1. Aristotele (Met.): Metá tá physiká theta, [Engl. trans. by S. Makin, Metaphisics Book Theta, Clarendon Press, Oxford 2006]
2. Blanchot, M.: The Space of Literature. University of Nebraska Press, Lincoln (1982) (1955)
3. Bollmer G.D.: Virtuality in systems of memory: toward an ontology of collective memory, ritual, and the technological. Memory Stud. **4**(4), 450–464 (2011)
4. Deleuze, G.: Difference and Repetition. Columbia University Press, New York (1994) (1968)
5. Deleuze, G.: The Logic of Sense. Columbia University Press, New York (1990) (1969)
6. Deleuze, G., Guattari, F.: Kafka: Toward a Minor Literature. University of Minnesota Press, Minneapolis (1986) (1975)
7. Deleuze, G., Guattari, F.: A thousand plateaus: Capitalism and Schizophrenia II. University of Minnesota Press, Minneapolis (1987) (1980)
8. Deleuze, G.: The Fold. The Athlone Press, Leibniz and the Baroque, London (1993) (1988)
9. Deleuze, G.: Essays Critical and Clinical. University of Minnesota Press, Minneapolis (1997) (1993)
10. Deleuze, G.: Immanence: a life. Theory, Culture and Society (1997) **14**(2) (1995)
11. Deleuze, G.: La soggettivazione. Corso su Michel Foucault (1985–1986/3), Verona: Ombre Corte (2020)
12. Deleuze, G.: On Painting Courses, March-June 1981. University of Minnesota Press, Minneapolis (2025)
13. De Souza, Silva, A., Suiko, D.: Theorizing locative technologies through philosophies of the virtual. Commun. Theory **21** (2011)

14. Didi-Huberman, G.: Ce que nou voyons, ce qui nous regarde. Gallimard, Paris (1992)
15. Firth, R., Robinson, A.: For the past yet to come: Utopian conceptions of time and becoming. Time Soc. **23**(3), 380–401 (2014)
16. Floridi, L.: The Fourth Revolution. Oxford University Press, How the infosphere is reshaping human reality, Oxford (2014)
17. Foucault, M.: The Order of Things: An Archaeology of the Human Sciences. Pantheon Books, New York (1970) (1966)
18. Joyce, J.: Ulysses. Vintage, H.W. Gabler (1986) (1922)
19. Kant, I.: Critique of Pure Reason. Cambridge University Press, Cambridge (1998) (1781)
20. Klee, P.: The Diaries of Paul Klee, 1898–1918. University of California Press, Berkeley (1964)
21. Kosari, M., Abbas, A.: Thirdspace: The Trialectics of the real, virtual and blended spaces. J. Cyberspace Stud. **2**, 163–185 (2018)
22. Lacan, J.: Encore, On Feminine Sexuality, the Limits of Love and Knowledge: The Seminar of Jacques Lacan, Book XX (1972–1973). Norton, New York (1998) (1975)
23. Lenco, P.: (Re-) Introducing Deleuze: new readings of Deleuze in International Studies. Millennium: J. Int. Stud. **43**(1), 124–144 (2014)
24. Lundborg, T.: The virtualization of security: Philosophies of capture and resistance in Baudrillard, Agamben and Deleuze. Secur. Dialogue **47**(3), 255–270 (2016)
25. Sarti, A., Montanari, F., Galofano, F.: Morphogenesis and Individuation. Springer, Heidelberg, New York, Dordrecht, London (2015)
26. Simondon, G.: On the Mode of Existence of Technical Objects. University of Minnesota Press, Minneapolis (2017) (1955)
27. Simondon, G.: Individuation in the Light of the Notions of Form and Information. University of Minnesota Press, Minneapolis (2020) (2005)
28. Simondon, G.: Imagination et invention. University of Minnesota Press, Minneapolis (2023) (2014)
29. Simondon, G.:, Sur la Tecnique. Press Universitaires de France, Paris (2014) (2014 b)
30. Smith, D.W.: Deleuze's concept of the virtual and the critique of the possible. J. Phil. Cross Discipl. Inq. **4**(9), 34–43 (2009)
31. Spinoza, B.: Ethics, Included in: The Collected Works of Spinoza, vol. 1. Princeton University Press, Princeton, 1985 (1677)
32. Thom, R.: Structural Stabulity of a General Theory of Models. W. A. Benjamin, New York (1975) (1972)
33. Thom, R.: Parables, Parabolas and Catastrophes: Conversations on Mathematics. Thombooks Press, Science and Philosophy (2014) (1983)
34. Sylvester, D.: Interviews with Francis Bacon. Thames and Hudson, London (1975)
35. Zanelli, S.: Con-fini. Deleuze, Simondon e il problema dell'individuazione. Castelvecchi, Roma (2023)
36. Zanzotto, A.: Fosfeni. Milano, Mondadori (1983)

Chapter 8
THALASSA

Elena De Silvestri

Abstract The essay offers a rereading of psychic regression as a generative movement rather than an involution, uncovering within it a possibility for subjective transformation. Drawing on Ferenczi's notion of "Thalassa," the return to the elemental, the fluid, the primordial ocean is interpreted not as an escape from reality but as a reactivation of a non-symbolized, sensitive dimension of the body. The desire for immersion in the aquatic expresses a need for disidentification from adult forms of subjectivity—a regressive gesture that becomes an opening to new modes of experience. The unconscious is conceived not as a closed structure, but as a topical and sonorous environment, echoing Fedida's insight: a space where the subject is exposed to the other and to the silence of origin. Regression, far from being pathological, reveals itself as a condition for metamorphosis.

The notion of regression, both in common usage and in some of its more specific meanings, seems to be characterized by a conservative and restorative tendency. In general, a regressive process describes a backward movement, from effect to cause, from the conditioned to the condition, from organized activity to less specified functional patterns. In any case, the meaning of regression depends on a progressive evolution, in relation to which it represents a process of decline or a literal repetition of what occurred at a lower stage.

The definition given in the psychoanalytic context does not seem to diverge much from this model: in *The language of psycho-analysis,* Laplanche and Pontalis write that "*Applied to a psychical process having a determinate course or evolution, 'regression' means a return from a point already reached to an earlier one.*" [25, p. 386]. In temporal terms, regression implies the existence of a genetic succession and denotes the subject's reversion to past phases of development (libidinal stages, object relations, identifications, etc.). From a *formal* point of view, it marks a shift to modes of expression and behavior located on a lower level with regard to complexity, structure, and differentiation. The order presupposed here is therefore

E. De Silvestri (✉)
Università di Verona, Verona, Italy
e-mail: elena.desilvestri@univr.it

S. Vizzardelli and R. Panattoni (eds.), *Plasticity and Discontinuity*, Lecture Notes in Morphogenesis, https://doi.org/10.1007/978-3-032-03119-8_8

not a succession of actually experienced stages, but a hierarchy of functions or structures. However, a third kind of regression exists. Initially, in fact, Sigmund Freud was not thinking in genetic terms, but in logical and spatial ones: here we speak of *topographical* regression, as in the case of dreams, in which the psychic apparatus takes a backward direction [14, p. 542][1] in relation to an oriented succession of virtually coexisting systems. So even though all these types of regression tend to coincide [14, p. 584], the notion of topographical regression—a return to a psychic place—appears marked by a certain eccentricity, thus undermining its dependence on evolutionary progress.

In the following pages, we will attempt a non-conservative reading of the notion of regression, beginning with the insights arising from a place, an environment: *thalassa*. The expression "thalassal regression" is used by the Hungarian psychoanalyst Sándor Ferenczi in the context of a work that undertakes a rather ambitious task: to develop a psycho-bioanalytic theory of genitality—modeled on what Freud had done in *Three Essays on the Theory of Sexuality* [15]—capable of encompassing both ontogenetic and phylogenetic perspectives. On the one hand, Ferenczi seeks to apply psychoanalytic models to the study of the physiology of tissues, organs, and their parts; on the other, to shed light on psychic phenomena through insights drawn from the field of biology. However, the aim here is not to validate or invalidate Ferenczi's bioanalytic approach, but rather to expand upon the intuition it raises at the metapsychological level. "Metapsychology," understood as a psychology that takes us "behind" consciousness—or a psychology of the unconscious—is indeed naturally characterized by a topographical regression, by a return outside of consciousness, where an unbridgeable rupture is inscribed. For this reason, describing and articulating the metapsychological field always requires rejecting any psychologistic formula and appealing to an *extraclinal* model. A sort of framework—or "fictional model," to borrow Freud's words—necessary for the convergence of two realms: the conscious and the unconscious, psychology and its beyond. Thalassa, or "the mode of existence abandoned in primeval times" is precisely one of these frameworks. It is an image able to grasp the living being in its most paradoxical relationship: that with the inorganic.

The hypothesis underpinning *Thalassa: A Theory of Genitality* appears controversial from the very beginning, also with reference to the analytical field. Ferenczi's way of using the metaphor of the ocean—particularly in reference to the so-called "oceanic feeling"—is not foreign to psychoanalysis or to Freud's thinking, but it is, indeed, a metaphor. Responding to Romain Rolland, Freud describes it as a tendency to feel at one with the world and with things, "a feeling as of something limitless, unbounded—as it were, 'oceanic'" [22, p. 64].[2] Specifically, this pervasive sense of

[1] "The only way in which we can describe what happens in hallucinatory dreams is by saying that the excitation moves in a backward direction. Instead of being transmitted towards the motor end of the apparatus it moves towards the sensory end and finally reaches the perceptual system. If we describe as 'progressive' the direction taken by psychical processes arising from the unconscious during waking life, then we may speak of dreams as having a 'regressive' character."

[2] See also the correspondence between Freud and Romain Rolland, who wrote to Freud about the "oceanic feeling" [21, letters 241, 242].

belonging is described as being linked to a disturbance in the ego's external boundaries. Far from being an original, autonomous, and unified condition, the ego is in fact, according to psychoanalytic theory, a particular "object" constructed to give the illusion of possessing these characteristics: "That such an appearance is deceptive, and that on the contrary the ego is continued inwards, without any sharp delimitation, into an unconscious mental entity which we designate as the id and for which it serves as a kind of façade—this was a discovery first made by psycho-analytic research, which should still have much more to tell us about the relation of the ego and the id. But towards the outside, at any rate, the ego seems to maintain clear and sharp lines of demarcation" [21, p. 66].

If the ego is intensively and metapsychologically layered inward, under normal conditions the boundary separating it from the external environment is clear and tangible. However, certain particular states of consciousness—such as being in love or experiencing pathological phenomena—violate this principle, establishing an ambiguous relationship with the surrounding environment and objects. Phenomena of this kind—including what we might call the oceanic feeling—would thus be linked to a regression that reaches back beyond the sense of self, still marked by a fusional relationship with the mother's body. In contrast to this structure, Ferenczi reverses the relationship: "the mother would, properly, be the symbol of and partial substitute for the sea, and not the other way about" [7, p. 54]. Going beyond the metaphor, the thalassal regression refers to "striving towards the aquatic mode of existence abandoned in primeval times" [7, p. 52] before the catastrophe of the desiccation of the waters took place. The author strives to uncover the way this phylogenetic catastrophe resonates in the mechanics of sexual drives. The individual relives this catastrophe—this time at the level of ontogenesis—in the trauma of birth. But beyond the more specifically bio-analytic hypotheses, what matters here is precisely the effect of this "reversed symbolism" by which the "force of attraction" exercised by Thalassa would be powerful enough to reverberate at both biological and symbolic levels.

Upon closer inspection, this shift seems to pursue a very specific goal: to find an escape route from the risk that Ferenczi defines as "psychomorphism," an image of the psyche that is too "psychic" or "psychologized". To better understand the meaning of this expression, it is helpful to go back to what the author sets out in the 1923 preface. Here Ferenczi explains the analogical method on which the entire structure of the text rests. First of all, his aim is to justify the way he brings together two worlds: the world of natural sciences (which investigate biological evolution) and the world of the human sciences (specifically the psychoanalytic approach). More generally, in Ferenczi's view, analogy seems to be the tool that can best question meaning. If, from a descriptive point of view—which identifies the phases or stages of a process—it is possible to limit oneself to the tools provided by a specific field of research, the derivation of meaning requires to cross boundaries. It demands an appeal to scientific domains foreign to the starting point. Thus, the physicist might resort to terminology from the human sciences to understand phenomena specific to his inquiry, while the psychoanalyst may search for the laws that govern the functioning of the psyche-system by looking at topical, dynamic, and economic metaphors [17,

p. 181].[3] Analogy therefore necessarily requires the criterion of *heterogeneity*: "Once the right is granted to make freer use of analogies previously despised, it is perfectly obvious that these should be drawn from fields as remote as possible. Analogies derived from related fields would tend to be, indeed, mere tautologies, and as such could hardly serve the function of proof. (…) Or, as a different approach, let us utilize the fact that materials are usually measured by a material of another kind; thus we easily may proceed to measure the material by the non-material, and vice versa" [7, p. 3].

From the point of view of meaning, the immaterial and the material thus each represent the measure of the other—like a spot that is not on the map that illuminates both the direction and the trajectory of the main road. All physical phenomena, Ferenczi continues, also demand a metaphysical explanation; conversely, all psychological phenomenon require a "metapsychological" framework. It is as if the term "metapsychology" here meant a return to a system of reference governed by the laws of physics.

However, the criterion of heterogeneity seems to reveal itself also in a second formulation, apparently similar to the first one, yet marking a more radical point. Its clearest expression might be found in an elliptical remark: beyond the survey of onto-phylogenetic stages, Ferenczi's aim is a "return to animism"—but an "animism no longer anthropomorphic" [7, p. 2]. An animism capable of deviating from the distinctly human connotation that the term "soul" tends to carry with it. In other words, a form of animation of the living that is neither anthropomorphic nor anthropomorphizable. In this sense, the use of analogy does not only serve a function of reciprocity, where two fields of knowledge, foreign to each other, assist one another. At a deeper level, what is inscribed here is an intensive collision point, where two separate realms meet in a restless synthesis. In other words, the analogical relation here is not merely dialectical, but seems intended to stun the fields of knowledge, to subject them to mutual infiltration, so that a minimal and paradoxical matrix may emerge: that of a psychic vertex that is not *psychomorphic*, an absolute foreignness internal to the living being.

We can thus better understand the reversal operated by Ferenczi between the geological element of the Ocean and the symbolic one of the mother. In a certain sense, it is precisely a matter of relegating the reference to the symbolic to a secondary place—or at least of making it *less human*, privileging an environmental referent. "To desire the ocean" means to orient a desire that is irreducible to identity schemas, a sort of environment-desire of a topographical nature, "made of qualities, substances, powers, and events" [3, p. 61].[4] In this sense, the world and experience are not

[3] "When we have succeeded in describing a psychical process in its dynamic, topographic, and economic aspects, we should speak of it as a metapsychological presentation."

[4] In this context, Deleuze criticizes the psychoanalytic framework that assigns to parental figures the role of central pivots of psychic experience. Such a perspective, in fact, fails to grasp desire in its original correlation with an environment: "The father and mother are not the coordinates of everything this is invested by the unconscious. There is never a moment when children are not already plunged into an actual milieu in which they are moving about, and in which the parents as persons simply play the roles of openers or closers of doors, guardians of thresholds, connectors or

interpreted through the Father-Mother pair, since parents themselves are places of immersion and passage. The mother is conceived here as a kind of threshold function: in the separation from the maternal body, the subject abandons its aquatic universe, just as the "retreat of the oceans" forced life into the trauma of becoming "terrestrial." The privileged position assigned to her derives from the fact that she is the one who grants and denies access to an oceanic stage of desire. But this reversed configuration has a second significant consequence. In this complicity with the material universe, desire—including sexual desire—necessarily takes on the meaning of a decomposition of symbolic identities. In coitus, Ferenczi writes, the regression exerted by Thalassa acts surreptitiously, producing "*the individual catastrophe of birth and the most recent catastrophe to the species, that of desiccation, but all the earlier catastrophes since life originated* as well; so that we have represented in the sensation of orgasm not only *the repose of the intrauterine state,* the tranquil existence in a more friendly environment, but also *the repose of the era before life originated,* in other words, *the deathlike repose of the inorganic world*"[5] [7, p. 63].

With thalassal regression, we are thus situated at the beginning of a double trajectory. In relation to the living being, a material environment is, on the one hand, *non-living*. The desire to return to matter implies a reflection on the desire for the extinction of life itself, as a coincidence with a life *before life*, or with the persistence of a non-anthropomorphic life at the heart of the living being. On the other hand, the ocean equally represents a composite space, traversed by movements in constant becoming, which bring forth a teeming dimension of form.

A distinction repeatedly used by Ferenczi may help clarify the convergence of these aspects. It concerns the description of two types of reaction that a living being can adopt in response to an event coming from the external world: the response can be *alloplastic*, leading to a reciprocal action on the surrounding reality, or *autoplastic*, triggering a withdrawal that results in self-modification. While alloplastic activity grants the ego a position of equilibrium, a stability in relation to external objects [9], in the autoplastic response, subjectivity emerges and is plastically altered in relation to a radical breakdown. Far from representing a defense mechanism, it intervenes precisely "when all defensive forces have been exhausted" and "the environment does not consent any longer to accept the role of being incorporated in this way" [10, p. 225]. Ferenczi defines it as a phenomenon of *materialization*, similar to what can be observed with hysterical conversion symptoms. It is a kind of material ecstasy, where an "altruistic pleasure"—the pleasure of yielding to another form—becomes indistinguishable from the mobilization of "impersonal elemental forces" [10, p. 225]. The example provided is that of a bird falling prey to a predator, which, fascinated by the sight of the snake or the claws of the eagle, after a short period of resistance, "throws itself to its own ruin." *Alloplastic* and *autoplastic* responses thus aim at two forms of stillness, both operating in the living being: the "egoistic"

---

disconnectors of zones. The parents always occupy a position in a world that is not derived from them" [3, p. 62].

[5] Here, Ferenczi makes explicit reference to the theses presented by Freud in *Beyond the Pleasure Principle* [20].

stillness of the ego, firmly rooted in a functional motor schema, is counterbalanced by another kind of stillness—"total" and "altruistic" stillness—in which the subject surrenders to the otherness of a foreign form. This is what occurs, for example, in the so-called *identification of the child with the aggressor*. According to Ferenczi, during childhood, identification is the response to an act that cannot be assimilated—one the child is unable to recognize or metabolize: an act of erotic seduction. Confronted with adult seduction, the infantile logic comes to a halt, generating a short-circuit between "languages": the language of the child and that of adults—the "language of passions," dramatically ambivalent, in which love and hate interpenetrate without any clear separation. Blocked in the execution of its natural automatisms, children "subordinate themselves like automata to the will of the aggressor, to divine each one of his desires and to gratify these; completely oblivious of themselves they identify themselves with the aggressor" [13, p. 162]. Instead of directing action outward toward the environment, in an alloplastic manner, they "only react in an autoplastic way by a kind of mimicry" [13, p. 163]. Ferenczi thus traces an essential link between oblivion and imitation, between the intrusion of a foreign behavior and the plasticity of the child's body. In this mimetic theater, the subject is less the director of a movement than its "receiver"—like a kind of recording plate, a sensitive body. Identification unfolds along a regressive vector, moving in the opposite direction of the incremental path of identity acquisition. Regression thus moves toward a de-subjectivized form, to the point that we could even speak of a death instinct—although "instead of 'death-instinct' it would be better to choose a word that would express the absolute passivity of this process" [11, p. 220]. The death instinct, reconfigured according to this tendency toward materialization, seems to have something to do with the passive capacity to receive an impression. On several occasions, Ferenczi emphasizes the importance of a fundamental function of the psyche, which consists in silencing one's body and 'being dead' (See [12]). Faced with an imminent threat, such an "anesthetic" function allows for the neutralization of bodily affections, at the cost, however, of developing a hypersensitivity to every kind of external modification. The alloplastic tendency is therefore, at once, the symptomatic response to trauma and a condition necessary to life itself in its relation to the environment: a "vital" symptom, which determines the morphological conditions of the living being, precisely insofar as it challenges its separation from the inorganic.

In *Des bienfaits de la dépression*: *éloge de la psychothérapie*, the psychoanalyst Pierre Fédida suggests interpreting in this direction the onto-phylogenetic perspective adopted by Freud and Ferenczi during the years of their closest correspondence. According to Fédida, the passage from biographical childhood to the childhood of the entire species—their coexistence along converging lines—is instrumental in accounting for the most *inhuman* manifestations of form, above all the teratological expression of the symptom.[6] If "ontogenesis recapitulates phylogeny"—according to Ernst Haeckel's formulation [23]—then trauma can no longer be reduced to being a mere accident in one's personal history, it is the entire history of humanity that

[6] The comparison with Ferenczi's theses on the topic of phylogenesis reappears in various places in the Freudian text (particularly in Freud [18, eng. trans. 1987]).

is in question here. Also, the symptom is not so much a pathological deviation, as "the essence of mental desease lies in a return to erlier states of affective life and of functioning" [19, p. 286]; see also: [15, p. 235], [16, p. 199];[7,8,9]). To take up a Freudian metaphor, neuroses offer a glimpse into a prehistoric landscape: symptoms are like fossils, life forms in which we can sense affective patterns and movements not yet culturally specified.

Fédida pays particular attention to glacial catastrophes, "the era of the renewed covering of large portions of the surface of the earth with ice and water" [8, p. 104].[10] The reference to the lasting marks left by the Ice Age is especially significant, as it offers a metapsychological landscape for those experiences of affective impoverishment in which life appears "inanimate." Foremost among these is *depression*: in depression, emotional life is significantly diminished, as if the body itself were bearing witness to an ongoing disappearance. Beyond the symptoms classically associated with the depressive state, Fédida's observations emphasize a specific trait, whereby the subject becomes the victim of an impersonal fascination—of the "enchantment exerted by a *dead state*, or by *a dead person*" [6, p. 46]. So much so that the subject's very appearance, their form, seems to signal a disappearance of the human *as such*.

Between the beginning and the end of the glaciation, it is indeed reasonable to suppose that animal and plant life forms devised autoplastic strategies to stay alive, minimizing as much as possible the expenditure imposed by external factors. What seems to have occurred, on a broad scale, is a stage of *depressivity*, in which biological forms appeared rigidified in a material ecstasy, which served to withstand the most violent and incompatible environmental conditions. What appears in the guise of a "glacial neurosis" [6, p. 39] in the pathological forms of contemporary life—the stillness of the body, which seems to be frozen, the low erotic and instinctual reactivity, the unresponsiveness to external stimuli—is in this sense a reactivation of this ancient past, which pertains to the origins of humanity. However, if the "depressive state" manifests an identitarian consolidation of the subject—identity as a state of death—which becomes inanimate in order to protect itself from the violence of life, what Fédida calls "depressivity" instead describes an intrinsic characteristic of the psyche. In fact, the author uses explicitly generic economic terms to describe it, referring to the mobilization and containment of conflicting psychic energies: "When I say that the depressive capacity is a characteristic trait of psychic life, I have in mind an 'economic' mode of life [...] at the origin of what is called the *psychic*" [6, p. 15]. What is designated, therefore, is a kind of negative faculty: that of *not* coinciding

---

[7] "The prehistory into which the dream-work leads us back is of two kinds—on the one hand, into the individual's prehistory, his childhood, and on the other, in so far as each individual somehow recapitulates in an abbreviated form the entire development of the human race, into phylogenetic prehistory too."

[8] "Ontogenesis may be regarded as a recapitulation of phylogenesis, in so far as the latter has not been modified by more recent experience. The phylogenetic disposition can be seen at work behind the ontogenetic process."

[9] On the influence of Darwin in Freud's work see [28].

[10] On the ice age in a psychoanalytic perspective see also [27].

with a "state," with a perfectly consolidated identity. Depressive economy always operates in relation to a limit against which it inscribes a counter-movement: when confronted with the subject's immobilization, it functions as a "reanimation of the living" [6, p. 83]; when confronted with the violence exerted by hyper-vitalism, it ensures a return to an "inanimate" form.

Beyond the scientific truth of these hypotheses, the ascent toward phylogenetic childhood thus has the advantage of isolating an ideal moment, in which the living form was oriented toward a kind of inorganic animism. In this analogical passage—in which the landscape of the soul is, in every respect, a geology—the psychic itself acquires an internal movement. Passing through the glaciation, life partakes in an existence close to *minerality* [4]; or rather, in a form of life that remains such on the condition that it is placed under the protection of death. As Fédida observes, "playing dead' is much more than pretending to be dead; more than preserving life, it means *producing the psychic as life*" [5, p. 22].

A famous essay by Roger Caillois devoted to mimetic phenomena seems to describe a remarkable convergence with these metapsychological theses. After a careful analysis of the classification of these phenomena, of the controversies surrounding their interpretation, and of the polymorphism of their occurrences, Caillois turns to the analogical method, invoking psychoanalysis and anthropology: "Beyond doubt some of the above developments are far from offering any guarantee from the standpoint of certainty. It may even seem questionable to compare such diverse realities as homomorphy and the external morphology of certain insects, sympathetic magic and the concrete behaviour of people of a certain type of civilization and perhaps a certain type of thought, and finally psychasthenia and the psychological postulations of people belonging, from these points of view, to opposite types. Such comparisons, however, seem to me not only legitimate (just as it is impossible to condemn comparative biology) but even indispensable as soon as we approach the obscure realm of unconscious determinations" [1, pp. 31–32].[11]

In the field of "unconscious determinations," the possibility of drawing analogies would therefore be not only legitimate but indispensable. However, Caillois is far from proposing a kind of "psychologization" of the phenomenon of mimicry; rather, the opposite is true, it is from the observation of mimicry that one can deduce something about the unconscious domain. In fact, it would be correct to say that here the unconscious is evoked to point out what inevitably escapes the rationale of identity, highlighting its "fallacious appearance," and forcing the introduction of a principle of discontinuity. Mimicry is, in fact, according to Caillois, a "pathology of difference," a visible demonstration capable of making one doubt the clear and distinct line by which the Ego persuades itself of being something different from its environment.

The observation of a mimetic body first of all forces one to question how one being *resembles* another. An "other" with which it shares no direct biological or hereditary relationship (whether it be a leaf, seed, bird, bark, or insect). There may be several hypotheses: similarity could be the flawed illusion of an anthropomorphic gaze, which perceives a good form where there is only, for example, the spherical charm

[11] On mimicry see also [2].

of the ocelli.[12] Or, resemblance could be the result of random factors, a combination like any other of certain particular elements that only accidentally came together in a way that results in a resemblance. Further, one could imagine a sort of "will" of the living being oriented toward the adaptation to a privileged object, driven by instinct and yet still imperfect in its functions. However, upon closer inspection, resemblance presents a problem of composition and decomposition: in what way has a form constructed itself bringing about that particular mimetic architecture? Or conversely, in what way has a supposed original form been decomposed, disarticulated, finally imitating something from the environment in which it lives? Caillois's hypothesis is, in every respect, regressive, according to a mode that is extremely instructive in this context. The author picks up on an observation by the biologist Félix Le Dantec about the Kallima butterfly, or orange oakleaf [26], whose morphology perfectly imitates that of a dry leaf. According to this hypothesis, it would be possible to imagine that the ancestors of the Kallima possessed cutaneous organs that permitted the simulation of the imperfections of leaves. This imitation mechanism, similar to a "sculpture-photography," would then have disappeared once the criterion of resemblance was achieved. Now, according to Caillois: "This hypothesis, which does not even seem to have been formulated, ultimately appears to be the only plausible one: it implies, in every respect, a certain plasticity of the organism at a given epoch, thanks to which its morphology was shaped by influences that are no longer operative today, at least under normal conditions" [1, p. 27].

At a certain point, and under very specific conditions, the organism exposed itself to a modification, interrupting its evolutionary movement and becoming singularly plastic, before "normal" life conditions were restored. Mimicry should therefore be defined as an "*incantation fixed at its culminating point*" [1, p. 27], a kind of dynamic fixation, entirely similar to the autoplastic reaction described by Ferenczi. This kinship is reinforced by the fact that Caillois names the pole of attraction of this plasticity as a form of "*temptation by space*" [1, p. 28].[13] The relationship of resemblance here is not one between two subjects or two objects. The other that one resembles is little more than an intermediary, an expedient, whereas the goal seems to be *assimilation into the environment*. The plasticity of form here is inseparable from the instinct to let go, to abandon, that polarizes being toward an inorganic mode of existence, and because in humans it is possible to detect psychological virtualities that are very similar, Caillois continues, we could define this mimetic attraction as a "legendary psychoasthenia"[14] [1, p. 28]. It is indeed a pathology of difference, a disorder of the boundary that separates the organism from its environment and likewise mobilizes virtual and autoplastic forces. Powers, qualities, substances that

[12] Lacan in *Seminar XI* takes up precisely these observations by Caillois, emphasizing the power of fascination exerted by the ocelli, which must be clearly distinguished from the Gestalt impression of a resemblance to the shape of an eye [24].

[13] see also Vizzardelli-De Filippis [29].

[14] The term "psychasthenia" refers to a type of psychoneurosis—a neurosis characterized by the predominance of psychological symptoms over physiological ones—essentially attributed to a weakening of psychic tension.

operate in the indistinction of this being and its environment and that only an *allo-plastic* activity—an activity complete and oriented toward the external world—is able to interrupt, isolating the former within the latter. Through mimicry, what becomes individualized is a regressive tendency of consciousness in general, whereby life itself regresses, takes a step back to rediscover the unconscious resources of the inorganic.

All these observations lead us to reconsider how the pair regression-fixation functions. These observations, in other words, underscore the ambiguous nature of the mechanism that links the movement of return with that to which one returns. If on the one hand fixation indicates a peak of permanence, of immobility, it is nonetheless a plastic immobility, an *incantation fixed at its culminating point*. Fédida effectively summarizes this conjuncture: "Like animal and plant forms, the psychic can be conceived as a fixed form of life preservation, though it is a form *plastic enough* to possess the capacity for regression" [6, p. 34]. Imagining an evolutionary movement does not simply mean establishing a time sequence in terms of before and after, as in a stage-based progression. Likewise, evolution includes these moments of "fixed evolution," one that is almost photographic. However, fixity is never actual, it never fully corresponds to the state of things in which it reveals its effects. When we say that a form regresses to its state of incantation, we must already affirm its dynamism, its mimetic plasticity. A relation of contamination that makes it permeable to the environment and whose mechanisms operate "beyond" what consciousness is able to grasp. Conversely, if the attractive force exerted by fixation is not functioning in the present and must be ascribed to the past, we should likewise say that it coexists in the present, in a state of virtuality. As if its ability to insist, under certain conditions, could always reactivate a regressive path.

Freud writes about fixation that "it takes place in a phase of development that was too strongly pronounced, or perhaps also that persisted so long as to pass entirely into the next phase" [18, p. 9]. Despite subsequent modifications and biographical events, the force of the impression is thus transmitted from stage to stage, bringing about, in the same form, a state of "co-existence": "Here succession also involves co-existence, although it is to the same materials that the whole series of transformations has applied" [19, p. 285]. Just as the insect, having reached perfect resemblance, renders the impression of an ancient image visible, fixation functions as a pole of internal virtualization. Here regression is articulated in the alternation of an actual and a virtual form, a coexistence that is destined to persist and resurface in all future transformations. To describe this quality, which refers as much to the past as to the future, Fédida speaks of "regressive progress" [6, p. 35].

For this reason, regression and fixation can only be conveyed by an ambivalent image. On the one hand, "mineralization" accentuates a kind of "zero degree" of form. It is at this level that we find the conditions allowing a form to maintain itself, to return to itself and discover the criteria of its own resemblance. As if the possibility of regression were what enables the subject to undergo a process of individuation that actualizes only some of the possible combinations of traits. But it is precisely within this immobility that the possibility of materializing something *other* is also maintained: the "fixed," "fossilized" form does not only indicate the persistence of a state, but also its disposition to "figure," to appear as a *figure*. So that the condition allowing the form to preserve itself as such is the same that announces its becoming-other, along virtual lines of development. Thalassa is an image that expresses this virtuality, and at the same time affirms its real and material nature, both living and inorganic. An image Caillois seems to perfectly describe when he writes: "[...] It becomes strikingly evident to what extent the living organism is one with the environment in which it lives. Around and within it, the same structures are present and the same laws operate. Thus, in truth, it is not merely *in* an environment, it *is* this 'environment', and the very energy that individuates it, the will of being to persevere in *its* being, is expended in the exaltation of itself and already secretly draws it toward the uniformity scandalized by its imperfect autonomy" [1, p. 32].

References

1. Caillois, R.: Mimicry and Legendary Psychasthenia. MIT Press, trans. John Shepley, October 31, 1984, pp. 16–32 (1935)
2. Caillois, R.: The Mask of Medusa. Clarckson N. Potter, New York (1964) (1960)
3. Deleuze, G.: What children say. In: Essays Critical and Clinical. Verso, London-New York (1998) (1993)
4. Fédida, P.: L'echo mineral. Musique en jeu, n. **9**, 116–118 (1972)
5. Fédida, P.: Le site de l'étranger. La situation psychanalytique. PUF, Paris (1995)
6. Fédida, P.: Des bienfaits de la dépression: éloge de la psychothérapie. Odile Jacob, Paris (2001)
7. Ferenczi, S.: Thalassa. A Theory of Genitality. Psychoanalytic Quarterly, New York (1968) (1926)
8. Ferenczi, S.: Male and female. In: Thalassa. A Theory of Genitality. Psychoanalytic Quarterly, New York (1968) (1929)
9. Ferenczi, S.: Autoplastic and alloplastic adaptation. In: Balint, M. (ed.) Final Contribution to the Problems and Methods of Psychoanalysis. Karnac, London (1994) (1930a)
10. Ferenczi, S.: Thoughts on 'pleasure in passivity'. In: Balint, M. (ed.) Final Contribution to the Problems and Methods of Psychoanalysis. Karnac, London (1994) (1930b)
11. Ferenczi, S.: Each adaptation is preceded by an inhibited attempt at splitting. In: Balint, M. (ed.) Final Contribution to the Problems and Methods of Psychoanalysis. Karnac, London (1994) (1930c)
12. Ferenczi, S.: The clinical diary of Sándor Ferenczi. In: Dupont, J. (ed.) Harvard University Press, London (1988) (1932)
13. Ferenczi, S.: Confusion of tongues between adults and the child. In: Balint, M. (ed.) Final Contribution to the Problems and Methods of Psychoanalysis. Karnac, London (1994) (1933)
14. Freud, S.: Regression. In: The standard edition of the complete psychological works of Sigmund Freud, vol. 5 (1900–1901): The interpretation of dreams (second part) and On dreams. The Hogarth Press and the Institute of Psycho-Analysis, London (1953) (1900)
15. Freud, S.: Three essays on the theory of sexuality. In: The standard edition of the complete psychological works of Sigmund Freud, vol. 7 (1901–1905): A case of hysteria, Three essays on sexuality and other works. The Hogarth Press and the Institute of Psycho-Analysis, London (1953) (1905)
16. Freud, S.: Archaic features and infantilism in dreams. In: The Standard Edition of the Complete Psychological Works of Sigmund Freud, vol. 15 (1915–1916): Introductory Lectures on Psycho-analysis (Parts I and II). The Hogarth Press and the Institute of Psycho-Analysis, London (1963) (1913)

17. Freud, S.: The Unconscious. In: The Standard Edition of the Complete Psychological Works of Sigmund Freud, vol. 16 (1914–1916): On the History of the Psycho-analytic Movement, Papers on Metapsychology and Other. The Hogarth Press and the Institute of Psycho-Analysis, London (1975) (1915a)
18. Freud, S.: A Phylogenetic Fantasy: Overview of the Transference Neuroses. Ilse Grubrich-Simitis (ed.) Harvard University Press, Cambridge (1987) (1915b)
19. Freud, S.: Thoughts for the times on war and death. In: The Standard Edition of the Complete Psychological Works of Sigmund Freud, Volume XIV (1914–1916): On the History of the Psycho-Analytic Movement, Papers on Metapsychology and Other Works. The Hogarth Press and the Institute of Psycho-Analysis, London (1975) (1915c)
20. Freud, S.: Beyond the pleasure principle. In: The Standard Edition of the Complete Psychological Works of Sigmund Freud, vol. 18 (1920–1922): Beyond the Pleasure Principle, Group Psychology and Other Works. The Hogarth Press and the Institute of Psycho-Analysis, London (1975) (1920)
21. Freud, S.: The letters of Sigmund Freud. In: Ernst, L. (ed.) Freud With a new Introduction by Steven Marcus (letters 241, 242). Basic Books, New York (1960) (1929)
22. Freud, S.: Civilization and its discontents. In: The Standard Edition of the Complete Psychological Works of Sigmund Freud, vol. 21 (1927–1931): The Future of an Illusion, Civilization and its Discontents, and Other Works. The Hogarth Press and the Institute of Psycho-Analysis, London (1961) (1930)
23. Haeckel, E.: Generelle Morphologie der Organismen. Reimer, G., Berlin vol. 2 (1866)
24. Lacan, J.: The four fundamental concepts of psychoanalysis: the seminar of Jacques Lacan Book XI (1964). Jacques-Alain Miller (ed.) Norton, New York and London (1998) (1973)
25. Laplanche, J., Pontalis, J.-M.: The Language of Psycho-analysis. Hogart Press, London (1973) (1967)
26. Le Dantec, F.: Lamarckiens et darwiniens : discussion de quelques théories sur la formation des espèces. Félix Alcan, Paris (1908)
27. Resnik, S.: Glacial Times: A Journey Through the World of Madness. Routledge, London (2005)
28. Ritvo, B.: Darwin's Influence on Freud: A Tale of Two Sciences. Yale University Press, New Haven and London (1990)
29. Vizzardelli, S., De Filippis, V.: La tentazione dello spazio, Orthotes, Naples-Salerno (2016)

MIX
Papier aus verantwortungsvollen Quellen
Paper from responsible sources
FSC® C105338

If you have any concerns about our products,
you can contact us on
ProductSafety@springernature.com

In case Publisher is established outside the EU,
the EU authorized representative is:
Springer Nature Customer Service Center GmbH
Europaplatz 3, 69115 Heidelberg, Germany

Printed by Libri Plureos GmbH
in Hamburg, Germany